BK 629.892 B496S
STATE-OF-THE-ART ROBOT CATALOG
1984 15.45 FV /BERGER, PH
3000 639818 30014
St. Louis Community College

W9-COT-834

THE STATE-OF-THE-ART ROBOT CATALOG

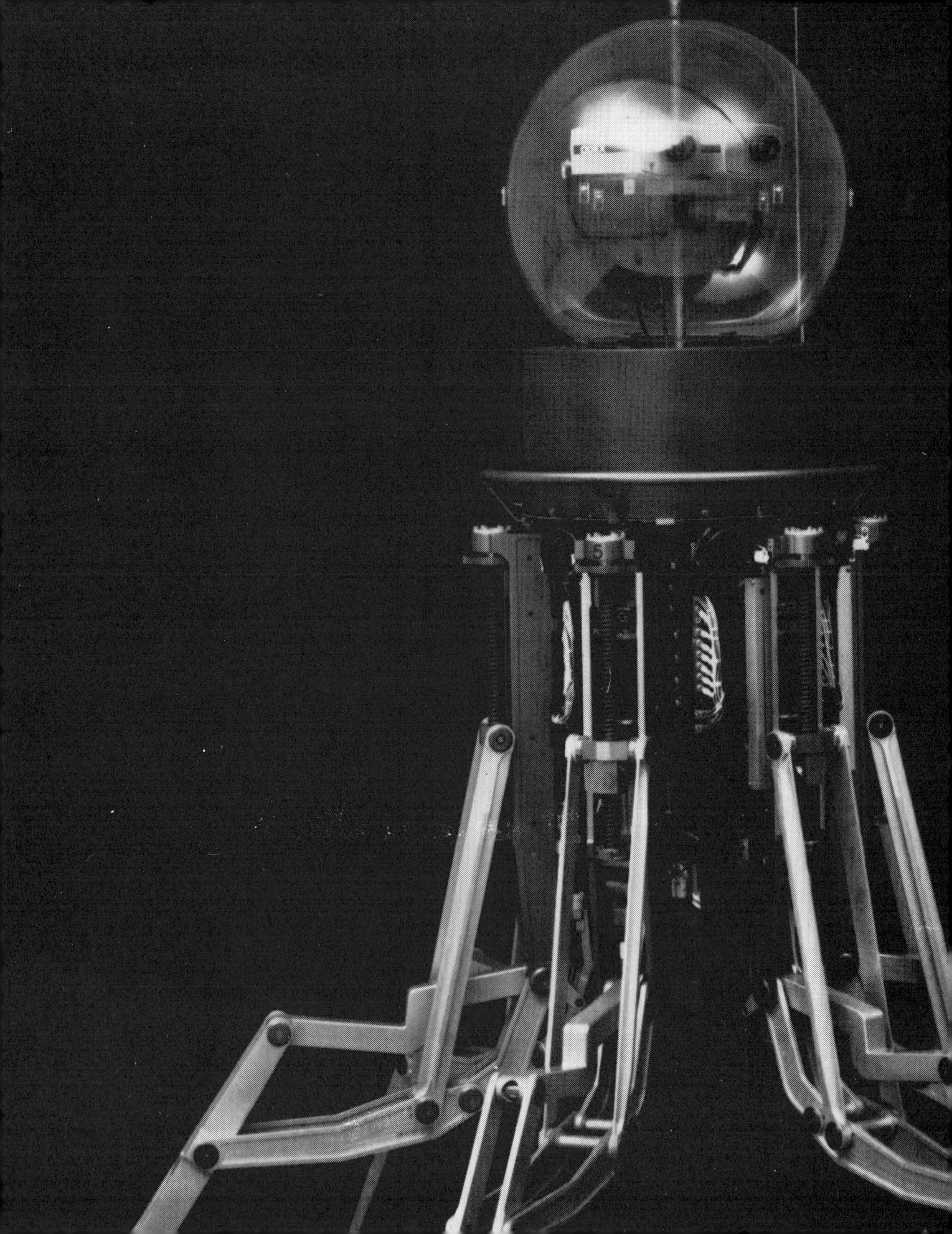

THE STATE-OF-THE-ART ROBOT CATALOG

Phil Berger

illustrated with photographs

DODD, MEAD & COMPANY New York

OTHER BOOKS BY PHIL BERGER

Miracle on 33rd Street: The N. Y. Knickerbockers' Championship Season
The Last Laugh: The World of the Stand-Up Comics
Deadly Kisses (a novel)

Published by Dodd, Mead & Company, Inc.
79 Madison Avenue, New York, N.Y. 10016
Distributed in Canada by
McClelland and Stewart Limited, Toronto
Manufactured in the United States of America
Designed by Stanley S. Drate
First Edition

Library of Congress Cataloging in Publication Data
Berger, Phil.
The state-of-the-art robot catalog.

1. Robots—Catalogs. 2. Robotics—History. I. Title.
TJ211.B48 1984 629.8'92 83-20838
ISBN 0-396-08361-7 (pbk.)

To Leslie,
with love

CONTENTS

ACKNOWLEDGMENTS

This book required the help and cooperation of literally hundreds of people. The following went above and beyond the call of duty:

Rene Miville; Maggie Steber; Jerry Bernstein of The Game Room; John Schilling of La Toya Lounge; Rabbi Zalman Kolko of Yeshiva Toras Chaim; Professor Edward Kafrissen of New York Institute of Technology; Ivan E. Sutherland of Carnegie-Mellon University; Les Goldberg of Odetix, Inc.; Donald Wollheim of DAW Books; Ann Flynn and Blair Whitton of the Margaret Woodbury Strong Museum; Veronica Chapman of Del Rey Books; Barbara Bazyn and Elizabeth Mitchell of Davis Publications; William Kenly of Paramount Pictures; Carl Flatow; Debbie Huglin of The Robotorium; Ronald Lynn of Bandai America Inc.; Leo Hoffman of Takara Toys; Joe Collins of World of Robots/Robotics International Corp.; Gene Beley of Android Amusement Corp.; Mary J. Bolner of The Robot Factory; Robert Doornick and Sandra Filippucci of International Robotics Inc.; Wil Anderson of ShowAmerica, Inc.; Bruce Barone of *Science Digest*; Sharon Smith of RB Robot Corp.; Lee A. Hart of Technical Micro Systems; Larry Smith of Sperry and Hutchinson; Robert Malone; Brian Wakefield of French and Rogers; Ellen Mohr of Unimation Inc.; Tom Akas and Sharilyn Shampine of the Society of Manufacturing Engineers; Lori Lachowicz of the Robot Institute of America; Thomas J. Keller of ASEA; William Werfelman of General Electric; Fanny and Jack Berger; Leslie Berger; Cynthia Darling; Kathryn Wardrop of GCA/Industrial Systems Group; Bob Wesley of Union Carbide Corp.; G. Scott Johnson of Exxon Company, U.S.A.; Sheila Evan-Tov; Ruth Dickey of NYU's Electronic Technical Aids Evaluation Project; Miriam Reid of Grumman Aerospace; Marc Kaplan of American Society for Technion; John Porter of Pizza Time Theater; Aleda Loinaz of Basic Telecommunications Corp.; Ellen Asher and Joe Miller of the Science Fiction Book Club; Rita Schreiber of *Robotics Today*.

My thanks to the following companies and institutes for their aid: Cincinnati Milacron; Mitsubishi Heavy Industries; The Franklin Institute; Musée d'Art et d'Histoire; Doubleday & Co.; Janus Films; Pyramid Film & Video; Colne Robotics, Inc.; Rhino Robots, Inc.; Androbots, Inc.; Heath Co.; Spine Robotics; Expert Automation; American Robot Corp.; International Robomation/Intelligence; Schrader Bellows; United States Robots; The DeVilbiss Co.; Comet Welding Systems; MTS Systems Corp.; Nordson Corporation; Control Automation Inc.; Intelledex Inc.; Automatix Inc.; Object Recognition Systems; General Motors Corp.; EWAB Engineering Inc.; Chrysler Corporation; NASA; Bell & Howell; Zymark Corp.; Ex-Cell-O Corp.; Veterans Administration.

A special thanks to my editor on this book, Jerry Gross.

NOTE

The prices and specifications of robots listed in this book were correct at the time this was written. Consult manufacturers for changes in specifications and price made since publication.

Photo references in parentheses in the text sections refer to photo captions and consumer information appearing in the catalog sections of this book.

INTRODUCTION

These are good times in robot industries, a period when increased public interest in robotics coincides with the growing commercial and intellectual intentions of the widely divergent group of men and women working to develop more sophisticated practical robots.

Out there are academics, engineers, businessmen, showmen, politicians, actors, authors, and laborers who, in one way or another, have a piece of a market whose time has come rather abruptly. What the computer was in the 1970s—a historic technological development—the robot has the potential of becoming in the 1980s.

The public is already cued to the robot phenomenon, its awareness keyed by the events and developments reported in newspapers, magazines, and on TV. Much of that media coverage concerns the novelty robots which are rolled out with regularity to publicize shopping malls and promote products, and the industrial robots—whose sweeping arms transport heavy materials or parts, or wield paint sprayers or welding torches—by now a familiar sight.

For purists, only the industrial robots count; that is, those robots conforming to the definition adopted by the Robot Institute of America in November 1979:

> A robot is a reprogrammable multifunctional manipulator designed to move materials, parts, tools, or specialized devices, through variable programmed motions for the performance of a variety of tasks.

For others, the definition provided by the *American Heritage Dictionary* will do:

> Any machine or device that works automatically, or by remote control.

Neither definition, though, addresses the gut-level way we humans respond to robots. At the Robots 7 Exposition held in Chicago in April 1983, amid displays of major robot companies (IBM, Unimation, Cincinnati Milacron), the Canadian town of Peterborough ("Our Views Are Today's . . . Our Business Is Tomorrow") had set up a booth that occupied a minuscule fraction of the square footage the industrial giants had. Yet Peterborough—attending the Exposition to attract industry to its city—did not lack traffic at its modest stand. Up against displays showing robots worth $100,000 or more, Peterborough still managed to draw crowds with a robot worth a pittance—a chess-playing robot as compelling as any of the high-priced hardware in other booths (1).

The patrons who watched the robot respond to chess moves made by flesh-and-blood opponents seemed to take inordinate pleasure in the ingenuity of the mechanical player. Never mind that technically it was less sophisticated than some of the whining, clanking robots that moved with great purpose in other corners of the exposition hall. What that robot board player managed to do was to touch the subconscious basis of our pleasure in robots. More than most of the expensive robots on hand at Chicago's McCormick Place East that day, the robot chess player's hook was most direct. For it could do what we could do. It was a fine line mechanical mimic of our own humankind, moving chess pieces as even you and I might. And, at bottom, that is what intrigues us about robots. In them we see a mechanical version of human life, a dance in steel that imitates our own steps. The people in Peterborough obviously knew something.

What they knew is speedily learned by any executive whose company deals in robots. James A. Baker, an executive at General Electric, touched on it in a presentation he gave to Society of Manufacturing Engineering Executives some time ago.

". . . Perspective," said Baker, "is very difficult to come by when the subject of robots arises. The newer robots are nothing more than computers with hands, but they produce a visceral fascination that draws and repels people far out of proportion to reality. People whose eyes glaze over [from boredom] at announcements of the most far-reaching electronics breakthroughs or business developments suddenly are gripped by the E. F. Hutton effect when someone mentions the word robot.

"I'll give you an example. In April [1981] GE held a press conference in New York and announced a massive commitment of company resources to a total war on the productivity problem, and our intention to develop, perfect, and market a wide range of automation systems that would eventually constitute what has been referred to as the 'factory of the future.' As part of our program we announced:

"The acquisition of Intersil Corporation—a pioneering integrated circuit company—for over a quarter billion dollars.

"The acquisition of Calma—a high technology computer-aided design and manufacturing firm—for $100 to $170 million, depending on sales performance.

"And the commitment to spend roughly another quarter billion dollars for R&D [research and development] focused on electronics in industrial applications.

"We also announced, in a short presentation, our entry into the robot business with the purchase of marketing and development rights for an assembly robot.

"The response from our audience of several hundred sophisticated financial analysts and press people was not exactly what we expected.

"In the exhibit hall our guests paid polite visits to our computer-aided design consoles, optical inspection systems, and other advanced industrial electronic equipment displays, and then made a beeline to another exhibit that was drawing more attention than Bo Derek on the beach—our assembly robot.

"The news coverage of the conference reflected the same fascination. Of seven major newspaper stories, five focused either totally or primarily on robots. The news stories on TV that evening ignored everything *but,* and one ended with the announcer speculating on whether TV anchormen could ever be replaced by robots. That particular feat, in many cases, could have been accomplished the day after television was invented, but the point is that robotic technology carries with it an anthropomorphic mystique with both negative and positive connotations that tends to distort its true role in industrial productivity."

At the Robots 7 Exposition in Chicago, exhibitors played to that anthropomorphic mystique. The man in the Peterborough booth, asked why his city would feature a chess-playing robot made elsewhere, said, with a sly smile: "To get people to stop here." Then, as he scanned the milling crowd at his booth, he added: "And it's working."

For the same reason, companies converted their robots from industrial tasks to more crowd-pleasing functions. A Swedish company, Spine, took its painting robot—a device that sweeps its long arm with fluid grace when it sprays—and had it dunk a sponge "basketball" through a hoop.

Visitors to Cincinnati Milacron's booth were given the opportunity to try their luck in a game of "21" that was run by the company's T^3-726 robot. The robot was equipped with Milacron's Milavision I Vision System, which enabled the T^3-726 to recognize the playing cards and handle them, and with a voice synthesizer by which the robot explained the rules of the game ("My cards, my rules," it said), told players their point totals, and announced winners and losers (2).

Such gimmickry is sophisticated play—and suggests the potential that robots possess for inventive applications. Just how far-reaching and crucial these applications will turn out to be is strictly guesswork at this stage. But the answer will surely determine the quality of our lives.

At this stage, robots are a rather new phenomenon, existing before the 1960s mostly in the popular imagination. Science fiction books and movies gave us robots that were entertainment artifacts—equivalents of horrific archetypes like Dracula and Frankenstein.

When technology at last caught up with pop notions, it came on the heels of a technological revolution that saw computer parts and prices shrink to a point where it now paid to deliver robots that could work or play.

As applications for robots broaden, technically adept laymen keep emerging who have the vision to utilize robots in novel ways. Like John Schilling, a restaurateur from West Boca Raton, Florida. Schilling's La Toya Lounge is noted for Willie, the Robot Waiter, who was created as a green furry frog. Willie gives the appearance of pedalling a bicycle-like cart topped by a serving tray (3).

"Willie," says Schilling, "is actually just a decoration for the real invention—the remote-controlled robot under the serving tray. That lower unit consists of two wheels and a body in between. Originally it was meant to be a piece of furniture placed next to a sofa or chair. On a voice command, it would roll to the kitchen, where the wife put a cup of coffee or whatever on it, and then roll back to its starting point. In other words, a kind of delivery cart.

"Then, in mid-1982, I took over La Toya Lounge here in West Boca Raton and decided to turn the unit into a mechanical waiter by setting a serving cart on top of the same two wheels, still giving me center pivot on turns for easy movability. I found, though, it was strange seeing the cart move around by itself. And that was where Willie, the frog, came into the program. He sits on the back of the cart and, while there are no mechanics inside of him, he improves the idea 100 percent. When Willie comes in, it looks as if he is peddling the cart. People go nuts."

The pleasure a robot waiter adds to the dining experience is reflected at La Toya's cash register. With Willie at work, Schilling's business has increased 15 percent, enough to give the owner ideas of selling La Toya so he can concentrate on other robot concepts. "I am," he confesses, "working on a new design for production. It will be simplified and streamlined—a series of carts. For instance, one will have a mailman mounted on it in much the way Willie is, and will be used for mail delivery in office buildings. Another will be—what else?—a waiter for restaurants. And still another will be a kind of Miss Piggy for factory parts delivery. They are all going to run on magnetic tape that is on the floor, so absolutely no operator will be required."

Through robots—and with a bit of ingenuity—the routine can be exalted. As David Moin wrote in *Long Island Jewish World*: "Our fondest recollections of our years in school are often of the times when the normal class schedule was broken by a school play, a special lecture, or perhaps a field trip. But there was never enough of this. School was largely predictable, fusty, and tied down to traditional modes of teaching. There is an Orthodox [Jewish] all boys school on Long Island, however, which seems to break from some of the traditions, and serves as a testament to the fact that learning can be fun."

The school to which Moin refers is Yeshiva Toras Chaim in Hewlett, Long Island. Rabbi Zalman Kolko, who has a background in engineering and teaches English and Hebrew subjects at the Yeshiva, has a robotic teaching aide. "Mr. Shamshi"—which is "my servant" in Hebrew—is five feet tall and programmed to ask students questions in subjects such as history and arithmetic, as well as Hebrew studies, and then to react to their answers. Given a correct response, Mr. Shamsi's nose lights up. To a wrong answer, the robot registers no response. The students? "They get so excited," says Rabbi Kolko (4).

What laymen like Schilling and Rabbi Kolko have done is to adapt existing technology to the daily routine. Hard-core robotics people envision even more radical applications. In Japan, it is predicted that the robot will be widely used by the end of the eighties on Japanese farms, in hospitals, and in dangerous primary industries such as mining.

According to a report of the Japan Industrial Robot Association: "If robots develop as planned there will be nurse robots to take care of physically handicapped people and aged patients in bed, to sweep the streets, to serve on fishing vessels, and to guide the blind."

Already the Japanese have tested "seeing-eye" robots as a replacement for seeing-eye dogs, and Mitsubishi has developed a remote-controlled underwater robot (5). In Israel, sighted robot fruit pickers are on the drawing board. And here in the U.S. the same quest for expanding the use of robotics goes on. As Tom Parrett, writing in *Science Digest,* notes: "Without question, formal robotics education will play a fundamental role in this growth. Already educators have taken note that future demand for skilled robotics technicians will be strong. At the New York Institute for Technology [NYIT] in Old Westbury, New York, for instance, Associate Professor Edward Kafrissen was asked two years ago to start a robotics laboratory. . . . The thrust of the lab is industrial robotic development, and Kafrissen's team has already built two prototypes." (6)

Through formal education—undergraduate* degrees in robotics and/or options in robotics as part of an engineering degree, are

*Graduate-level courses are given at such schools as Carnegie-Mellon University, the University of Houston, the University of Illinois at Chicago, Duke University. Drexel University, Georgia Institute of Technology, and Rensselaer Polytechnic Institute.

presently offered at universities like Duke, Oregon State, Carnegie-Mellon, Miami (Fla.), Florida, South Carolina—or through the sort of Yankee ingenuity that individuals like John Schilling and Rabbi Kolko have shown, we can anticipate great strides being made in what is essentially a new field. Even as you read this, there are men working on robotic creations that at one time would have seemed a fiction writer's fancy. For example, the walking robot that Ivan E. Sutherland is developing at Carnegie-Mellon University in Pittsburgh—a six-legged creature controlled by a built-in microcomputer which now requires a human driver. Some day, Sutherland hopes, the walking robot may function entirely on its own (7).

Another intriguing multilegged device is Odex I, which is being created by Odetics, Inc. of Anaheim, California. The manufacturer says: "Odex I is a multifunctional, multilegged walking robot with unprecedented strength to weight ratios and agility. The functionoid is designed to walk and work on any terrain. The feasibility model is comprised of six articulators, or legs, attached to a primary structure. The self-contained power source is located at the bottom of the primary structure, while the command/control center is on top. Above the command/control center is a stable platform for payload transport." (8)

Odex I is powered by a standard 24-volt, 25 AH aircraft battery, with 360 watt hours. This model can run for approximately one hour on 350 watts. Odex I is currently operated by radio control, and can adjust its height and width to the terrain it must traverse. According to the newsletter *Industrial Robots International*: "Work is proceeding on addition of vision and other sensory devices which, together with a computer program, would give it complete autonomy. . . . The basic unit as now being shown would sell for $200,000."

Applications for Odex I, according to the manufacturer, include mining, agriculture, space, sea, and land exploration, security, surveillance, forestry, construction, material handling, nuclear power plants and utilities, fire fighting and prevention, medicine, and military.

For those who find their imagination fired by the varied activity in robotics, and want to be part of the coming revolution, or for those who simply wish to know what's state of the art, this book serves as a guide to the myriad possibilities, providing the sort of basic detailed information—manufacturers' names, addresses, phone numbers, and prices—that will enable readers to get as involved with robotics as they choose to be. Whether it is a science fiction novel that features robots, or a hands-on experience with a working robot, the information for seizing these and other robotic opportunities is here in *The State-of-the-Art Robot Catalog*.

CATALOG AND CONSUMER INFORMATION

(1) Here is the chess-playing robot that the Canadian city of Peterborough used to draw crowds to its booth at the Chicago Exposition. Chess-playing robots are available from The Game Room, 2130 Broadway, New York, New York 10023. Tel.: (212) 595-0923. *Photo by Phil Berger.*

CONSUMER INFORMATION:
The Game Room's best-selling chess robot is The Sensory Chess Challenger 9, manufactured by Fidelity Electronics. "The Sensory Chess Challenger 9," says the store's owner, Jerry Bernstein, "sells six times better than all other robot chess players. That's because it combines low prices [$165] with what we call 'high strength,' which is a reference to the computer's chess ability by U.S. Chess Federation strength standards. Challenger 9 is 1770 strength: it will beat 90 percent of the players who compete with it. The Challenger 9 operates on sensory input: lights light up in the squares that the computer indicates moves to and from. You then move the piece by hand.

"A more novel chess robot is put out by Milton Bradley and is called The Grand Master [price: $399]. Here the piece will move on its own through magnetic contact. The chess pieces are much smaller than the squares, which allows them to move freely. The Grand Master is rated about 1600 strength. Frankly, many chess players prefer other chess robots that are more portable [The Grand Master must be plugged in] and are cheaper.

"I think the best chess robot for 1984 will be a new model that Fidelity Electronics is bringing out: The New Elite. Price is $389 and it's 1900-plus strength. It's smarter and has better aesthetics. Also moves by sensory input."

1

The Game Room gives demonstrations of its chess-playing robots for potential customers and will take phone orders. Bernstein accepts American Express, MasterCard, and Visa. Add $10 for shipping charges.

(2) Cincinnati Milacron's T^3-726 used vision and voice synthesis to serve as dealer in a friendly game of "21" that ran continuously at one of its booths at Robots 7. Two show visitors at a time were able to try their luck against the robot. *Photo courtesy of Cincinnati Milacron.*

CONSUMER INFORMATION:
Other T^3-726 robots at the Chicago exposition performed more traditional industrial tasks. The T^3-726 is for parts handling, assembly, and welding applications. Price: $65,000 (fully equipped). Interested parties contact Cincinnati Milacron, Industrial Robot Division, 215 S. West Street, Lebanon, Ohio 45036. Tel.: (513) 932-4400.

(3) Willie, the Robot Waiter, is on the job at John Schilling's La Toya Lounge. The robotic elements are in the lower unit between the two large wheels. *Photo by Michael Berger.*

CONSUMER INFORMATION:
Schilling says: "Each wheel has its own 150-pound pull torque motor, which can be operated by memory or remote control. With one wheel forward and one in reverse, the unit turns on center pivot for minimum turning space." La Toya is located on U.S. 441 & Marina Blvd., West Boca Raton, Florida 33433. Tel.: (305) 426-2200.

(4) Rabbi Zalman Kolko built "Mr. Shamshi," a robotic teaching aide for his students at Yeshiva Toras Chaim in Hewlett, New York. *Photo courtesy of Yeshiva Toras Chaim at South Shore.*

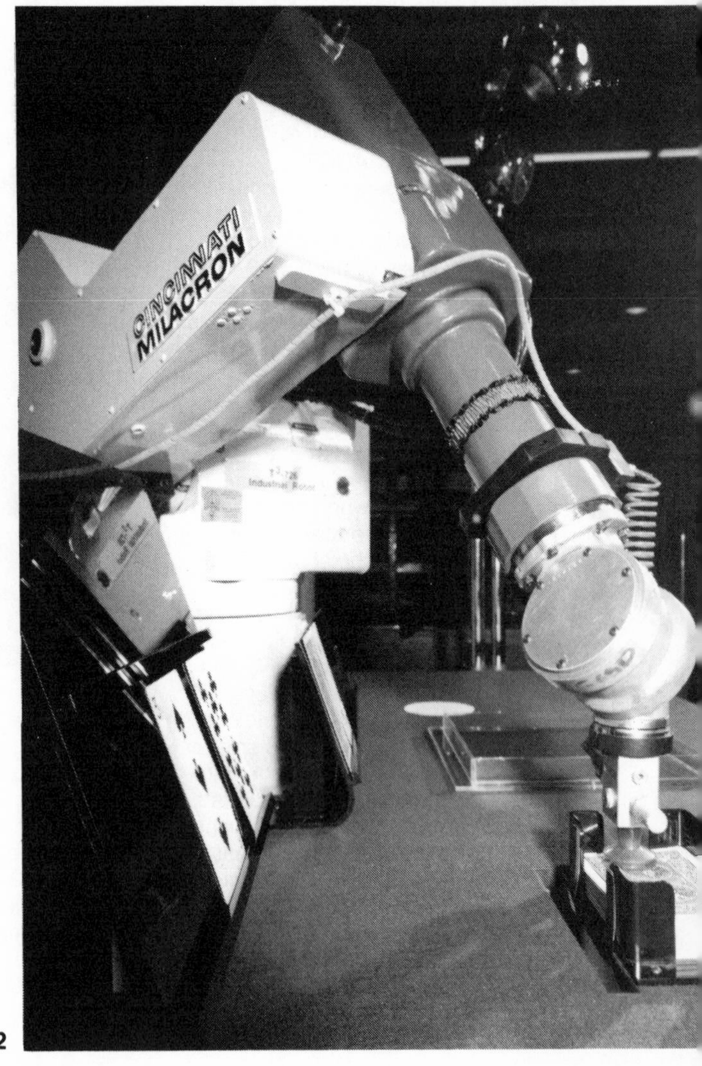

2

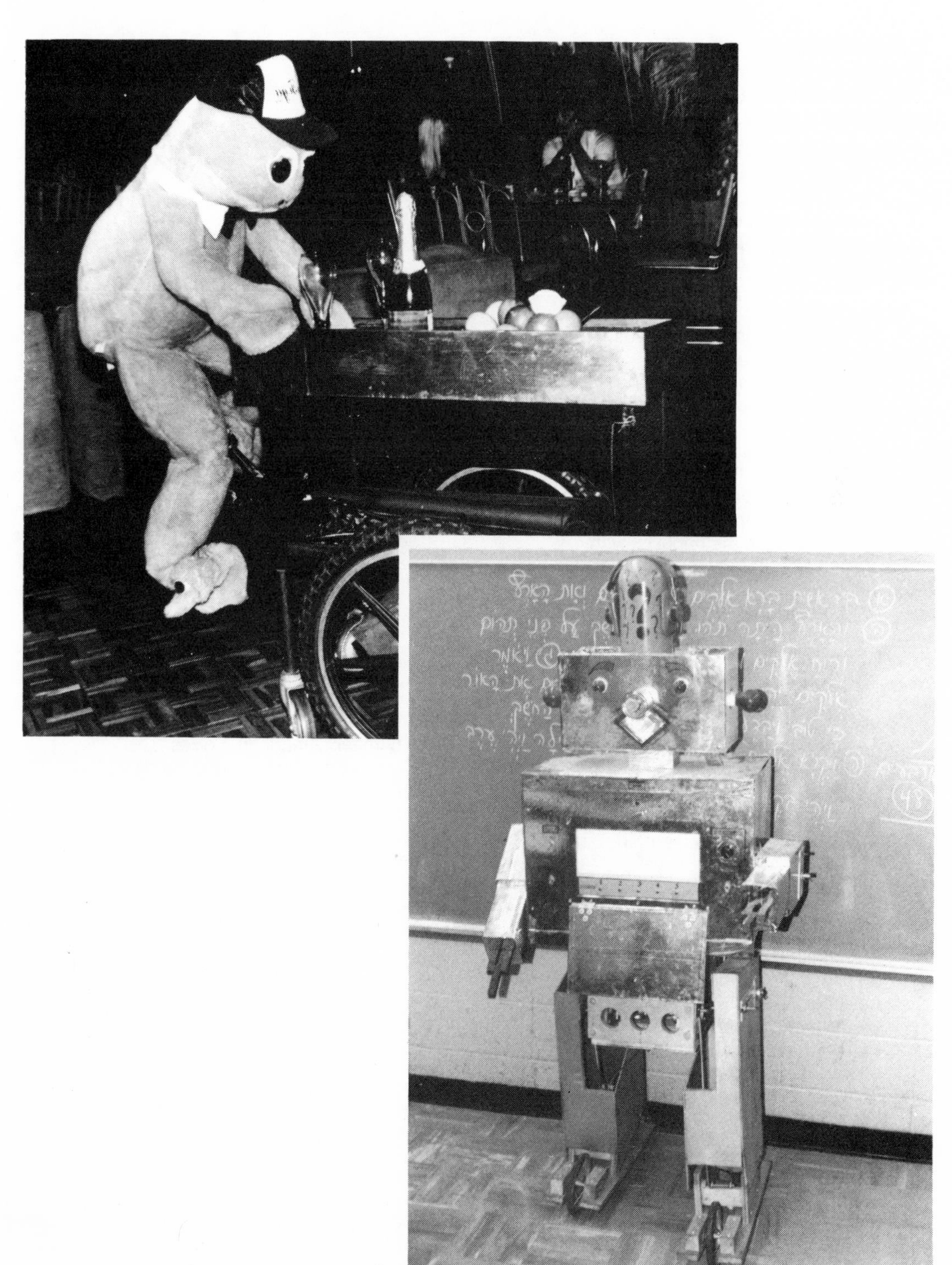

4

CONSUMER INFORMATION:
For hobbyists interested in building their own robot, RB Robot Corporation offers several books that would be helpful:

(R101) *How to Build a Computer-Controlled Robot,* by Tod Loofbourrow. $8.60.
The book provides detailed directions for building a computer-controlled robot with a KIM-1 microprocessor. Construction steps are pictured and described with clear and complete control programs, photographs, diagrams, and tables.
(R106) *The Complete Handbook of Robotics,* by Edward L. Safford, Jr. $9.95.
How to design and build any type of robot, including those with microprocessor brains. Data on interfacing robots with computers and on radio controls provide the hobbyist with expanded opportunities for experimentation.
(R107) *How to Build Your Own Working Robot Pet,* by Frank DeCosta. $7.95.
Shows you how to build your own robot dog, and program its IQ and temperament. With an 8085 microprocessor brain, this dog fulfills the dual functions of a good robot and a good pet.
(R110) *Build Your Own Working Robot,* by David L. Heiserman. $6.95.
Step-by-step instructions on how to build a robot with many human characteristics from readily available electronic parts. This book is about an unusual robot named Buster, who has been specially designed for amateur experimenters who want to try their hand at robotics. Buster has some basic reflex mechanisms, a will of his own, and even a personality of sorts. Includes plans, schematics, logic circuits, and wiring diagrams.

To order books, send for an order blank to: RB Robot Corporation, 14618 W. 6th Avenue, Suite 201, Golden, Colorado 80401, Attention: Book Orders. RB accepts check, money order (payable to RB Robot Corporation), or credit card account number (American Express, MasterCard, Visa). Postage and handling: an additional cost of $1.50. Colorado residents add 3 percent sales tax. For orders outside the USA shipping costs are added C.O.D.

5

(5) The underwater robot, from Mitsubishi. *Photo courtesy of Mitsubishi Heavy Industries, Ltd.*

CONSUMER INFORMATION:
Mitsubishi has developed an underwater robot to clean marine life (such as mussels, barnacles, bryozoa and/or algae) from pipes used in thermal and nuclear power generation plants. In the past, cleaning such pipes required drying them out in their entire length, or asking divers to be involved. According to the manufacturer, it was a "very unpleasant and filthy job with bad smell." Mitsubishi claims no antifouling paint, coating, or treatment is

6

required; sea water is not polluted at all. For further information, contact Mitsubishi Heavy Industries, Ltd, Power Systems Headquarters, 5-1, Marunouchi 2-chome, Chiyoda-ku, Tokyo, Japan. Tel.: Tokyo (03) 212-3111.

(6) Research and development in robotics is taking place in such colleges as New York Institute of Technology in Old Westbury, New York. Here, one of NYIT's student assistants (who are generally paid) is seen in the school's laboratory, which is run by Professor Edward Kafrissen. *Photo courtesy of New York Institute of Technology.*

7

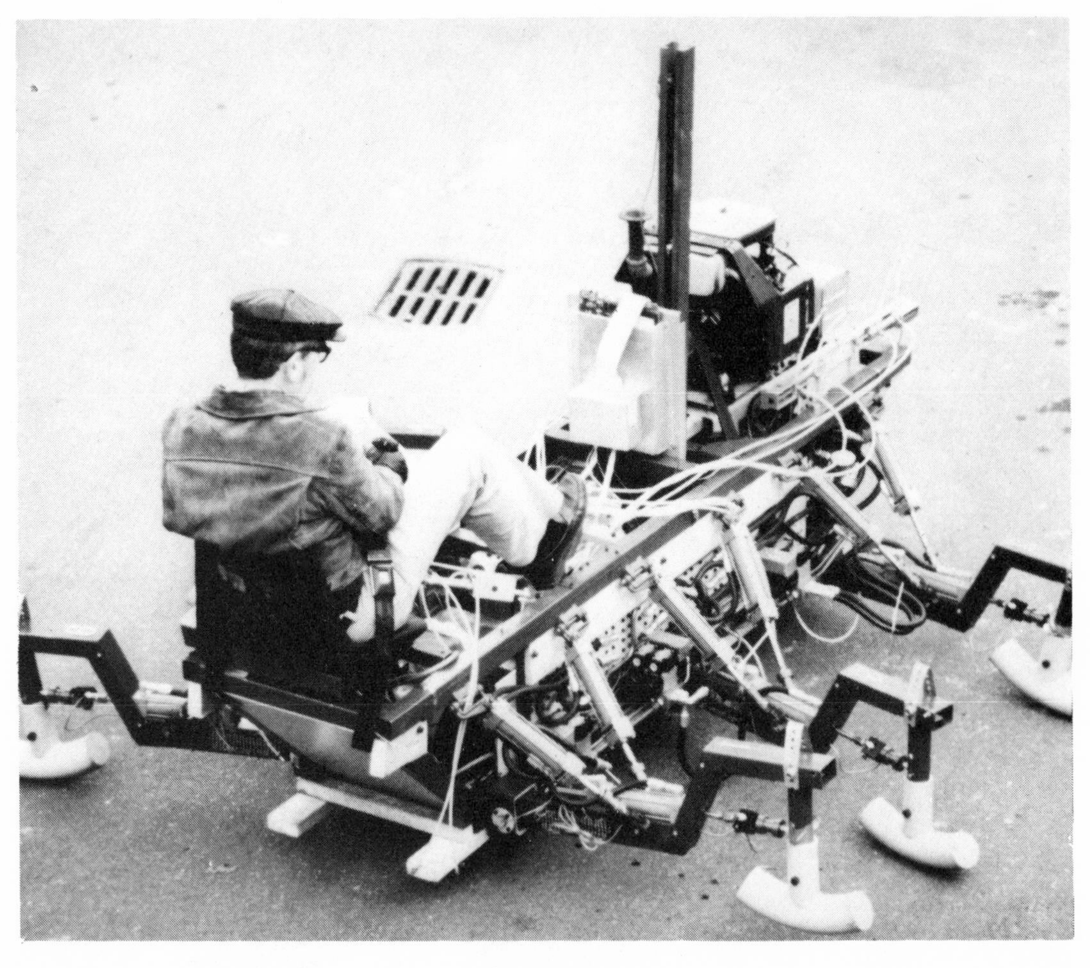

(7) This is Ivan Sutherland's walking robot. *Photo courtesy of Ivan E. Sutherland.*

CONSUMER INFORMATION:
Sutherland has written a book about his device, entitled *A Walking Robot.* To order, write to Marcian Chronicles, PO Box 10209, Pittsburgh, Pennsylvania 15232. Price is $25, plus $2.00 for shipping first book (50 cents for each additional book). Quantity discount for five or more books is 30 percent. Make checks payable to Marcian Chronicles, Inc. Pennsylvania customers include 6 percent Pennsylvania sales tax. Marcian Chronicles accepts payment by MasterCard, Visa, and American Express. Write for order forms.

(8) Odex I, a feasibility model. *Photo courtesy of Odetics, Inc.*

CONSUMER INFORMATION:
The load capacity of any one articulator of Odex I would be 450 pounds while the unit is operating, or 600 pounds stationary. For further information, contact Odetics, Inc, 1380 S. Anaheim Blvd., Anaheim, California 92805. Tel.: (714) 774-5000.

8

MYTHOLOGICAL ROBOTS

Robots from history, stage, screen, and literature

MYTHOLOGICAL ROBOTS

Though the word *robot* is fairly new to modern man's vocabulary, the idea of contrivances that simulate his motions is not.

As early as the fourth century B.C., Aristotle wrote: "If every instrument could accomplish its own work, obeying or anticipating the will of others . . . if the shuttle could weave, and the pick touch the lyre, without a hand to guide them, chief workmen would not need servants, nor masters slaves."

Aristotle's words described machinery that was what we now call automated, and at the time was a long way from being a reality.

Aristotle's vision coexisted with mythology of human life being created from inanimate substances, as in the story of the Greek god Hephaestus, who brought forth two living female statues from pure gold. Similarly, the mythical artist Pygmalion forged the statue of a woman whose effect on him was so overpowering that he beseeched the gods to imbue it with life.

History shows that the Greeks also had practical ideas about robot-like devices. In 400 B.C., Archytas of Tarentum is said to have made a wooden pigeon that could fly. And in the second century B.C., Heron of Alexandria used mechanical figures to demonstrate hydraulic, mechanical, and pneumatic action.

Nor were the Greeks alone in their preoccupation with figures that mimicked lifelike movement. In the fifteenth century, the German astronomer Regiomontanus reportedly made a fly of iron that would circle a room before returning to his hand, and also fashioned a mechanical eagle that flew when he exhibited it in the presence of the Emperor Maximilian.

The Renaissance brought elaborate clock towers with mechanical figures that struck the hour. Some of these clocks exist even today. In Piazza San Marco in Venice, two giants still hammer the tower bell on the hour as they have done since 1497. In Nuremberg's Frankenkirche there is a glockenspiel on which figures bring in the hour with a flurry of motion set to music. Other famous tower clocks exist at Berne and Messina and in the Cathedral at Strasbourg.

By the eighteenth century, the expertise that forged the Industrial Revolution would lead to the creation of automata—self-moving machines in which the principle of motion was contained within the mechanisms themselves. As robotics expert Robert Malone writes: "It became possible to create, within a very small space, machinery capable of controlling a whole sequence of actions. With the machinery and power source so compact, the new automatons, if life size, could be made much more complex; simpler ones could be reduced to miniatures. Aided by craftsmen who made clocks, watches, and dolls, the automaton maker brought his own ingenious solutions and meticulous craftsmanship to the simulation of life."

The automata from that period are still with us, scattered in museums throughout the world. And they are no less remarkable for the technological progress we have made since the 1700s. The automata created by Pierre Jacquet-Droz and his son Henri-Louis are particularly renowned.

Pierre Jacquet-Droz (1721–1790) was a master Swiss clockmaker with a reputation throughout Europe for quality. Prospering at his trade, he had the time—and the resources—to divert his skills to the creation of automata. Three such mechanized puppets by Pierre and his son Henri, first shown in 1774, are on display today at the Musée d'Art et d'Histoire in Neuchatel, Switzerland.

In their time the automata of the Jacquet-Drozes were widely admired, and were shown in the drawing rooms of high society and in the royal courts of Europe. In 1878, the automata were sold. They passed from owner to owner until, in 1906, they were bought by the Historical and Archeological Society of the Canton of Neuchatel for 75,000 francs. Not long after, they were installed at the Musée d'Art et d'Histoire, where they are activated for the general public on the first Sunday of every month.

Among the Jacquet-Droz automata that attract museum goers is "The Scribe" (9)—the figure of a male child seated on a stool, its right hand gripping a goose quill. So subtle are its mechanics that when the automaton begins to write his eyes actually follow the flow of the letters. The mechanism can be set so that it will write any text of forty or fewer letters. Of the three existing Jacquet-Droz automata, the mechanism (10) of this one is the most complex. From the museum's guide to the Jacquet-Droz mechanical puppets:

Pierre Jacquet-Droz must have been faced with very difficult problems: the major one being how to lodge the entire mechanism inside this child-size body and how to make the elbows and the arms command the movements of the wrists.

There are two distinct sets of wheel works. An ingenious system sets them alternately in motion until, without any interruption, the last full stop has been accomplished, thereby automatically bringing the whole machinery to a halt.

The first mechanism is situated in the upper half of the body. It propels a long cylinder on a vertical axis composed of three sets of cams, each of which controls the levers which, in turn, direct the movement of the Scribe's wrist in the three fundamental directions.

In this way the quill not only travels on a single plane but is equally capable of tracing the upstrokes and downstrokes of each letter. Then the second mechanism starts: it conveys to the cylinder an upward or downward movement of translation. The length of this stroke is determined by a disc situated in the lower half of the movement; this disc has forty interchangeable steel pegs fixed on its periphery, each peg being set at an angle of nine degrees. Each individual peg impels the cylinder into a determined position corresponding to a specific letter or change of gesture (i.e., beginning a new line, dipping the pen in ink, etc.).

Although it bears a startling facial resemblance to Pierre Jacquet-Droz's "The Scribe," "The Draughtsman" (11) was made by his son Henri. It took Henri two years to build his "Draughtsman," but when finished it was capable of rendering detailed sketches—a portrait of Louis XV, for instance. During those moments when the sets of cams change their position, the figure blows the dust off his sketching pad, accomplishing this through a bellows placed in its head. With three interchangeable sets of cams, "The Draughtsman" can make four drawings.

Another of Henri Jacquet-Droz's automata reflects his interest in music. Henri, a part-time composer, built the figure of a young girl seated at an organ-like instrument. "The Musician" (12) actually presses the keys of the organ, playing five harmonized motifs thought to have been written by Henri. While she plays, her eyes and head move to the sound of the instrument and, at the final notes of the song, "The Musician" executes a graceful little bow.

The automata of the Jacquet-Drozes inspired other craftsmen of the time. One of them, Henri Maillardet, was an apprentice under Jacquet-Droz and, during his life, created several pieces, the most famous being the "Writing Child," which was built in 1805 and is now the property of the Franklin Institute in Philadelphia (13).

When museum officials there acquired the automaton, it had not worked in years and had been damaged by fire. No one at the museum knew exactly how old it was or who had made it. The staff spent several months restoring the figure's delicate mechanism. When, at last, the mechanism at the base of the figure was activated, "Writing Child" produced the following verse in French:

A child, the darling of the fair,
Throughout the world as I dare tell
The ladies love me everywhere
And so their husbands do as well.

For that feat alone applause was merited. But the poem was not all that the automaton managed. In its very next line it did what museum personnel could not: it identified its creator with these words: "Written by Maillardet's automaton."

Maillardet's "Writing Child" weighs about 250 pounds and features a figure kneeling before a writing desk. Originally, the figure had been of a little boy in court dress. By the time the Franklin Institute acquired the automaton, its costume was that of a French soldier. In its restoration by the museum, it was fitted with a dress instead of a boy's suit and that is how it exists today.

"Writing Child" can write and draw seven pieces, including a detailed sketch of an early sailing vessel. The one English poem she can produce reads as follows:

Unerring is my hand, tho' small.
May I not add with truth:
I do my best to please you all;
Encourage, then, my youth.

In their time, the technically complex devices of Maillardet were sold to the wealthy and, on occasion, were even commissioned as state gifts. King George III reportedly gave the Emperor of China a mechanized Maillardet doll that was programmed to write flattering words about the Emperor in Chinese.

History records at least one episode of fraud involving the automata of this period. In 1769, Baron Wolfgang von Kempelen created a Turkish figure who sat before a chessboard, holding a pipe and responding to the moves of human opponents with moves of his own that repeatedly checkmated them. For many years, von Kempelen's famous figure astonished and puzzled Europe. In time, though, the figure was exposed as a hoax, ingeniously concealing the real performer—a dwarf who was hidden by a trick door.

The vogue for these lavishly produced automata passed, and soon production was a more cut-rate scale. Items ended being sold not to royalty but to the general populace. Many of these congenial objects are on view today at the Margaret Woodbury Strong Museum in Rochester, New York. Although the museum's collection does not include any of the most famous eighteenth-century automata, it does contain representatives of almost every type of mechanical doll and toy from the nineteenth and twentieth centuries. They range from a highly complex clockwork figure with nineteen separate movements, to the mass-produced, spring-wound tin toys that flooded the market after World War I.

Some of the museum's items are considered valuable. "The Burning Building," (14) says Blair Whitton, curator of toys at the Strong Museum, "is very rare. And so is the one of Ulysses S. Grant (15). You're talking about toys that go for five thousand dollars and up . . . depending on their condition . . . at auction."

"The Burning Building" is not, strictly speaking, an automaton. Activated when a cord is pulled, it features a fireman who scales a ladder, swings the maiden in distress over the balcony and down to safety, while another fireman trains water from a hydrant onto the flames.

The automaton called "General Grant" is seated on a cast-iron chair atop a wood box base. This figure of Ulysses S. Grant inhales and exhales smoke, and his arm moves the cigar to and from his mouth. General Grant is one of more than 100 automata at the Strong Museum. Other automata include a tearful child weeping over her broken doll; a girl enjoying a chocolate bar while she reads a book; a woman powdering her nose. There are also foreign-made pieces like "The Waltzers" (16) and "The Gay Violinist" (17). "Living Picture," a clockwork scene (18), is also on display.

In a technologically advanced society like ours it is hard to imagine the effect that automata had on people in earlier days. Artifacts like the Jacquet-Droz figures were especially celebrated. Audiences were drawn to them, in the words of one observer of the period, "as on a pilgrimage."

Among those who viewed the Jacquet-Droz automata in 1816 was a writer named Mary Wollstonecraft Shelley. She was the second wife of the English poet Percy Bysshe Shelley and she also became the author of *Frankenstein,* published in 1818, only two years after she had seen the Jacquet-Droz exhibit at Neuchatel. Whether the Jacquet-Droz display inspired the Frankenstein tale is moot. Less fuzzy is the position that Frankenstein occupies in the literary history of robots. Frankenstein—though created from human parts—is nonetheless an integral piece of the robot lore. Frankenstein was by his unnatural condition a terror to his species, a half-involuntary criminal, and finally an outcast whose sole recourse is self-immolation.

The dark note that Mrs. Shelley sounded would echo across a

century and, in 1921, when the concept of the robot as a mechanical version of man—a tin Frankenstein, if you will—was offered in a theatrical production called *R.U.R.*, the vision behind the play had the same ominous tone of *Frankenstein*. *R.U.R.* was notable in one other respect: it brought the word *robot* into the English language (19).

Of Czech origin, *robot* is derived from the word "robit" meaning "to work," and is related to an old Slav term "robota," meaning servile labor. The Czech playwright Karel Capek (pronounced CHOP-ek) created the word for his fantasy drama, *R.U.R.*, an abbreviation for Rossum's Universal Robots, the fictitious manufacturer of the robots depicted by Capek. *R.U.R.* was presented first in Prague and a year later, in 1922, was brought to New York. It was revived by the Theatre Guild in 1929.

The robots in Capek's play were lifelike creatures that were produced in volume and intended to replace human workers. They were mechanically efficient but lacked all feeling. Eventually, these robots rebelled against their masters, sparing a lone man to create more of them. But the formula for producing robots had vanished with the victims of their mayhem. In the end, though, one robot couple capable of feeling for each other leave the audience with the glimmer of hope that new beginnings are possible.

In spite of that optimistic prospect, the overpowering impression that *R.U.R.* created was of the dire consequences that awaited humans if they did not harness the technology at their command. The robot became the horrific symbol of a future in jeopardy. The vision suited Capek, who had been affected by the bloody efficiency he saw in the technology of World War I—a technology that brought mass death and destruction.

R.U.R. was a seminal work on robots and set the tone for the perception of robots by most authors and movie makers in the years that followed the play's earliest productions. The taboo suggested by *Frankenstein* was carried on by Rossum's Universal Robots: attempting to fashion artificial life was tampering with nature. Man had no business doing that. Not only was such activity outside his province, it risked upsetting the universal order, leading to chaos.

Such was the conventional wisdom of many creative people, who found in the robot a symbol that fit with their vague uneasiness about the technology that was transforming society. In the decade in which Capek wrote *R.U.R.*, commercial air travel came of age, the family automobile began rolling off assembly lines in unprecedented numbers, and radio brought us in closer contact. It was a pivotal time in our history. Change was in the wind, and where that change would take us made many people fearful.

Tied as that change was to technology—and given the bloody

consequences of modern armaments in the so-called Great War just concluded—it was natural for intellectuals to be dubious about the direction our society was taking. That doubt was reflected in how the robot—another step in a continuing mythology that went back to the ancient Greeks—was used in our fantasy life.

Science fiction pulp magazines were on newsstands shortly after World War I. Hugo Gernsback's *Amazing Stories* was the pioneering publication. But it was with John Campbell's *Astounding Stories** that science fiction took root. The early issues of the pulps often featured robots that were manlike in design—that is, they had arms, legs, and recognizable faces. In many cases, they also had a capacity for trouble. The portentous echo of *Frankenstein* and *R.U.R.* played through these tales.

> The robot was harmless, in fact couldn't be moved in any way whatever— til a newsman stuck his nose in where it didn't belong!

was the blurb on "Farewell to the Master," written by Harry Bates in the October 1940 *Astounding* (20). And:

> Gallagher, the mad, or at least pie-eyed scientist, had produced a remarkable robot. But seemingly useless. It spent its time admiring its unquestionably remarkable, if not beautiful, self. But Gallagher had to find out—but quick—why he'd made the infernal thing . . .

was the blurb for Lewis Padgett's "The Proud Robot" in the October 1943 *Astounding*, a tale about a seemingly unflappable robot that suddenly develops the potential for catastrophe (21).

That motif—the latent danger in the robot's apparently smooth works—was a familiar one dished up by pulp writers, but a variation was occasionally tried, with the robot depicted as capable of serving man without any hitches.

The writer credited with bringing us this three-dimensional view of the robot is the prolific Isaac Asimov. In the early 1940s, Asimov—a regular contributor to *Astounding*—began writing robot stories (22) which included a robot psychiatrist, Dr. Susan Calvin. She not only helped robots with their thought processes, but also tried to cure mankind of its distrust of mechanical help.

Asimov also created his Three Laws of Robotics—rules of behavior that are programmed into his robots' positronic brains and which act

*Through the years, the magazine was also referred to as *Astounding Science Fiction* and *Astounding*.

as a brake against the sort of dark impulses that Capek's robots had. The Three Laws:

1. A robot may not injure a human being or, through inaction, allow a human being to come to harm.
2. A robot must obey the orders given it by human beings except where such order would conflict with the First Law.
3. A robot must protect its own existence as long as such protection does not conflict with the First or Second Law.

In 1950, Asimov collected nine of his robot tales—written between 1940 and 1950—into a book entitled *I, Robot* (23). And while the complexity of Asimov's robots affected later writers' view of the mythological figure, it did not end a literary obsession with the dark side of these fantasy creatures.

Robots were now, like their flesh-and-blood creators, possessed of heroic and villainous possibilities and shades between. Sometimes those possibilities stretched the notion of merely mechanical help, as the jacket copy of Tanith Lee's novel, *The Silver Metal Lover,* suggests:

> He seemed utterly natural and perfectly human. He had total coordination and beautiful grace. He was handsome: eyes like two russet stars, eyelashes and hair dark cinnamon. Skin pale silver.
>
> He was a robot, and he could do everything a man could do—yes, everything. . . .
>
> To Jane he was real. *Real.* He made her all she could be, and some things she thought she couldn't, and she would do anything not to lose him.
>
> But to society he was an omen of ultimate threat—a man who excelled men in every way. And society knew how to deal with threats.
>
> What would *you* do if the manufacturer decided to recall the machine who just happened to be your one true love?

These days the robot of our imagination is a science fiction staple, seen by writers from a variety of angles (24-26). As the lunar probes of the pulps in time were equaled by events, so we have found curious echoes from robot fiction in our day-to-day reality.

Philip K. Dick in his novel, *We Can Build You* (27), wrote of a manufacturer that built exact simulacra of famous men. From the jacket copy: "They thought that people would pay a good price to have anyone they wanted made to order—to talk with or to utilize. . . ."

In Tokyo, Japan, Shunichi Mizuno, president of Cybot Co. Ltd., is constructing android robots. For twenty years he has worked to make his robots humanlike—eight years simply to develop the skin. His first success is a computerized Marilyn Monroe robot that uses a microcomputer-controlled air compressor to regulate facial expressions (28).

The movies in which robots appear show the same artistic progression as those in literary fiction. In the industry's earliest depictions, like the 1927 silent film *Metropolis* (29-30), robots were often a source of violence and disruption. That trait does not disappear even when, like Asimov's robots, some of Hollywood's mechanical men take on sunnier personalities, like Tobor the Great (31) in the Republic film of the same name (1954) and Robby the Robot (32) in MGM's *Forbidden Planet* (1956), and Gort (33) in Twentieth-Century Fox's *The Day the Earth Stood Still* (1951).

On the silver screen, for every robot bad guy like Gog (34), or the robot gunfighter in *Westworld* (35), or the silver-faced Law Robots (36) of *THX 1138,* there were robots who meant well like Gort, or Robby who, though intellectually superior to humans, were governed by a set of rules similar to Asimov's Three Laws. Hollywood came to find myriad possibilities in robots. In *Logan's Run,* a robot named Box (37) quick-freezes escapees from a futuristic domed city controlled by an omnipotent computer. In *Heartbeeps,* there is a robot comedian named Catskill (voice by Jack Carter). In *Silent Running,* there are three small robotic drones—Huey, Dewey, and Louey (after the Disney cartoon ducks)—who play poker with the film's star, Bruce Dern. From rape—*Demon Seed (38)*—to comedy—*Sleeper* (39)—to the ongoing heroics of R2D2 and C-3PO (40) in the *Star Wars* trilogy, robots are now thought to possess diverse possibilities.

That is fitting. For the robot that was once pure myth no longer is. And though today's industrial robot—regarded as crucial in the economy of the future—looks nothing like the humanized objects of the science fiction pulps or Hollywood films—man is still inclined to anthropomorphize. A film short called *Ballet Robotique* (41), which was nominated for an Oscar in 1983 in the category of Live Action Short Film, shows robots found at various General Motors plants across the United States. But it does not show them in their objective reality. As producer director Bob Rogers told writer Buck O. Shaw in *American Cinematographer*: "Each type of robot has its own personality. One moves like a chicken. Another like a cat. One paint-spraying robot moves like seaweed, gracefully swaying and curling with the current. Our objective was to use the cinematographer's art to enhance and emphasize each of these personalities."

Rogers embellished the robots' factory environments through special lighting and other technical aids. In the case of a chassis paint-sprayer robot, he gave the device a pair of painted red eyes that glowed in an atmosphere lit like an underwater cavern. Each time paint shot from the nozzle, the robot would appear to be snooping along the chassis, looking for areas on which to unload its paint. "By dramatizing the personality of each robot," Rogers told the magazine, "we're extending an old animation tradition."

As for the tradition of mythological robots, that surely has changed. In March 1949, the cover of *Astounding Science Fiction* (42) depicted a robot with a paper heart pinned to its metallic chest—a forward-looking vision of a machine that possessed the potential to be more than a tin hulk. The Alejandro illustration seemed to anticipate today's robotic age at a time when the vogue was to view robots as clanking perils.

That historically deep-rooted vision of robots as a threat to mankind, and their creation as a violation of nature, has been fundamentally altered by the rapid developments in the robotics field. The refined capabilities of today's industrial robots, as well as the emerging new personal robots, have thrust us into an age where myth is superseded by the daily science. With progress has come a projected tomorrow in which the old dark mythological sentiments give way to a more hopeful mood, a mood nicely captured by photographer Carl Flatow in a shot of his (43) that eventually became the April 1983 cover of *Omni Magazine* (44). Flatow's striking color picture (black and white in this book) shows a human hand reaching toward a silvery robotic hand in a gesture of a new day's trust and alliance.

Catalog and Consumer Information

(9-12)
These automata are the works of Jacquet-Droz father and son: **(9)** "The Scribe," by Pierre Jacquet-Droz. **(10)** A look at the inner workings of "The Scribe." Pierre's son Henri created **(11)** "The Draughtsman" and **(12)** "The Musician." All were first exhibited in 1774. *Photos courtesy of Musée d'Art et d'Histoire, Neuchatel, Switzerland.*

CONSUMER INFORMATION:
The Jacquet-Droz automata are currently on display at the Musée d'Art et d'Histoire in Neuchatel, Switzerland. The museum is located at 2 Rue des Deaux, Neuchatel, Switzerland. Tel.: (038) 25 17 40. Museum hours: from 10 to noon and from 2 to 5 Tuesday through Sunday. The Jacquet-Droz automata are activated for the public the first Sunday of every month.

9

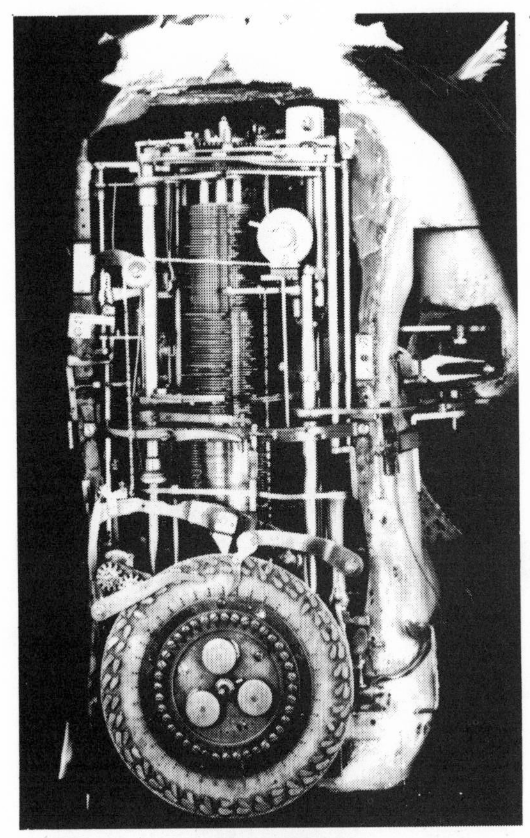

10

12

11

(13) The "Writing Child," an automaton by Henri Maillardet. *Photo courtesy of the Franklin Institute.*

CONSUMER INFORMATION:
The "Writing Child" is on display at the Franklin Institute, 20th & The Parkway, Philadelphia, Pennsylvania 19103. Tel.: (215) 488-1000. Museum hours: 10 to 5 Monday to Saturday; 12 to 5 Sunday. Admission: $3.50 for adults; $2.50 for children. Prints of "Writing Child" are available in the Institute's gift shop.

(14 & 15)
Two of the more valuable mechanical toys on exhibit at the Margaret Woodbury Strong Museum: **(14)** "The Burning Building," a pull toy made of cast iron and wood, about 1895 and **(15)** "General Grant," a clockwork toy (automaton) made of cast iron and wood, about 1880. *Photos courtesy of the Margaret Woodbury Strong Museum.*

CONSUMER INFORMATION:
The Margaret Woodbury Strong Museum is located at One Manhattan Square, Rochester, New York 14607. Tel.: (716) 263-2700. Museum hours: 10 to 5 Tuesday to Saturday; 1 to 5 Sunday. Closed Mondays. Admission: $2 for adults; $1.50 for students with school ID and for senior citizens; 75 cents for children ages 4 to 16; children under 4 (accompanied by an adult), and members are admitted free. Membership fees range from $35 to $250. Two video monitors showing toys being activated are on continuous display. Twice daily—1:30 and 2:30—there are live displays of toys being activated.

13

14

15

(16-18)
Also on display at the Strong Museum:

(16) "The Waltzers," a clockwork toy (automaton), about 1885. Manufacturer: G. Vichy, France. Materials: bisque heads, metal base. Two dolls in dancing position. As music plays, they revolve and move along a three-wheel base, changing directions.

(17) "The Gay Violinist," a wind-up toy, about 1910. Manufacturer: Martin Fernand, Paris, France. Made of tin, wire, lead, and cloth. When wound, the feet of the violinist vibrate, the figure moves about, with right arm easing bow along violin.

(18) "Living Picture," a clockwork scene, about 1900. Large living picture shows a shoemaker and helper. When clockwork is activated, shoemaker's arms move as he sews sole on shoe, his head wagging from side to side. Boy helper taps shoe with hammer, his head bobbing and foot kicking the cobbler's bench. Woman figure emerges from behind door with shoe clenched in hand. *Photos courtesy of The Margaret Woodbury Strong Museum.*

16

17

18

(19) Scene from the Theater Guild production of Karel Capek's *R.U.R.* A footnote: Spencer Tracy played the part of a robot in one of his first stage performances in an early 1920s production. *Photo courtesy of The Billy Rose Theater Collection; the New York Public Library at Lincoln Center; Astor, Lenox, and Tilden Foundations.*

CONSUMER INFORMATION:
Photos and memorabilia of *R.U.R.* are on file for public examination in The Theater Collection at The Research Center of the Performing Arts, third floor of the Lincoln Center Library, 111 Amsterdam Avenue, New York, N.Y. 10023. Tel.: (212) 870-1639. Open to public Monday to Saturday 12:00 to 5:45.

Photos of *R.U.R.* may be purchased at the Research Center of the Performing Arts. The 8 × 10 black and white glossy below costs $6.25.

(20) Traditional nemesis view of robots is seen in Harry Bates's short story, "Farewell to the Master" (illustration for "Farewell to the Master" by Harry Bates, October 1940) in *Astounding,* Copyright 1940 by Street & Smith Publications, Inc.; renewed 1967 by The Condé Nast Publications, Inc; used by permission of Davis Publications, Inc. *Photo by Phil Berger.*

CONSUMER INFORMATION:
Old copies of early science fiction pulps like *Astounding* are available at Fantasy Archives, 71 Eighth Avenue, New York, N.Y. Tel.: (212) 929-5391. Eric Kramer of Fantasy Archives suggests that collectors write to him, stating what issues they are seeking. Prices, Kramer says, vary according to the condition and age of the magazines.

19

20

21

REASON

By Isaac Asimov

The robot was strictly logical, reasoning, as only its perfect machine mind could, from observed facts to inevitable—if wacky—conclusion.

22

(21) Illustration for "The Proud Robot" by Lewis Padgett; October 1943 *Astounding,* Copyright 1943 by Street and Smith Publications, Inc.; renewed 1970 by The Condé Nast Publications, Inc.; used by permission of Davis Publications, Inc. *Photo by Phil Berger.*

(22) Isaac Asimov brought a more enlightened view of robots to science fiction, through short stories like "Reason." Illustration for "Reason" by Isaac Asimov, April 1941 *Astounding,* Copyright 1941 by Street & Smith Publications, Inc.; renewed 1970 by The Condé Nast Publications, Inc; used by permission of Davis Publications, Inc. *Photo by Phil Berger.*

(23) Asimov's *I, Robot* is a science fiction classic—a collection of nine of his robot short stories, written over a decade beginning in 1940. First published in hardcover by Gnome Press, 1950. *Photo by Phil Berger, by permission of Doubleday & Co., Inc.*

CONSUMER INFORMATION:
According to Eric Kramer of Fantasy Archives—a bookstore specializing in science fiction materials—a mint copy of a hardcover first edition of *I, Robot* is valued at $300. He occasionally has such a copy for sale. Kramer operates at 71 Eighth Avenue, New York, New York (one flight above street level). Tel.: (212) 929-5391.

Doubleday & Co., Inc. has brought out in hardcover *The Complete Robot* by Asimov ($19.95) which includes all the robot stories from *I, Robot,* plus others. *I, Robot* is also available in paperback (Fawcett, $2.50).

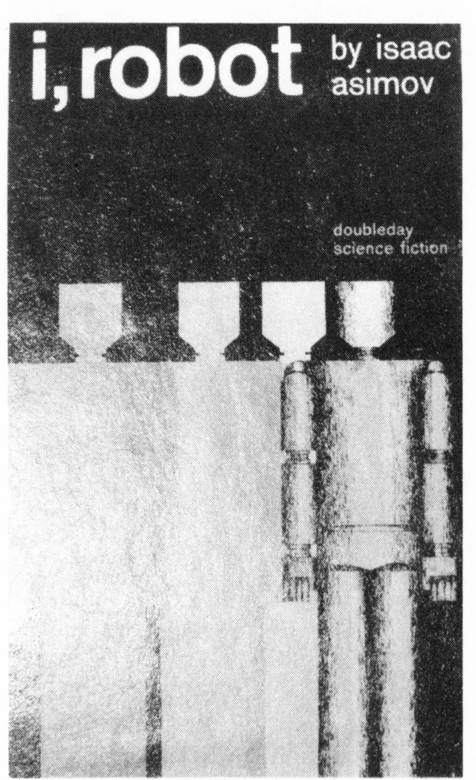

23

(24-26)
Robots have become a staple in science fiction, as the covers of these Del Rey Books indicate. *Photos by Bob Stamper.*

(24) Illustrated by Ralph McQuarrie, from the book entitled *The Best of Jack Williamson,* a Del Rey Book published by Ballantine Books; Copyright © 1978 by Jack Williamson; Reprinted by permission of Ballantine Books, A Division of Random House, Inc.

(25) Illustrated by Michael Whelan, from the book entitled *Special Deliverance,* a Del Rey Book published by Ballantine Books; Copyright © 1982 by Clifford D. Simak; Reprinted by permission of Ballantine Books, A Division of Random House, Inc.

(26) Illustrated by David B. Mattingly, from the book entitled *Code of the Lifemaker,* a Del Rey Book published by Ballantine Books; Copyright © 1983 by James P. Hogan; Reprinted by permission of Ballantine Books, A Division of Random House, Inc.

CLASSIC SCIENCE FICTION
THE BEST OF
JACK WILLIAMSON
INTRODUCTION BY FREDERIK POHL
REDONDO BLVD

24

A Charming Tale of Adventure and Survival by
CLIFFORD D. SIMAK
A DEL REY BOOK
SF
Ballantine
29140/$2.75
SPECIAL
DELIVERANCE

25

a novel by
James P. Hogan
CODE OF THE
LIFE MAKER

26

(27-28)
Sometimes science fact can be as beguiling as science fiction. The story line of Philip K. Dick's *We Can Build You* is based on the manufacture of celebrity robots **(27).** In Japan today, a Marilyn Monroe robot **(28)** is leased for promotional purposes. Photo of *We Can Build You* by Bob Stamper, courtesy of DAW Books, 1633 Broadway, NYC. Photo of Monroe robot by UPI.

CONSUMER INFORMATION:
The life-sized robot of Marilyn Monroe can mouth a song, shrug her shoulders, wink an eye, and strum a guitar. Japanese electrical engineer Shunichi Mizuno created the robot and leases it for advertising and display for approximately $10,000 a month, according to published reports. Mizuno can be reached at Cybot Co. Ltd, Tokyo. Tel.: (03) 262-9337.

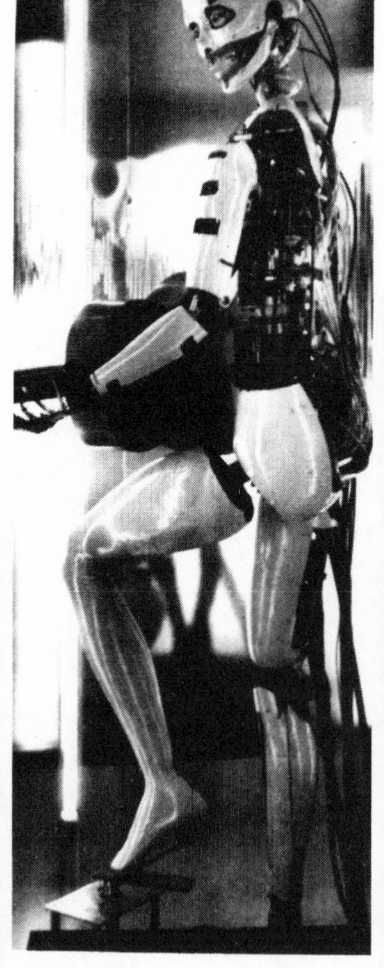

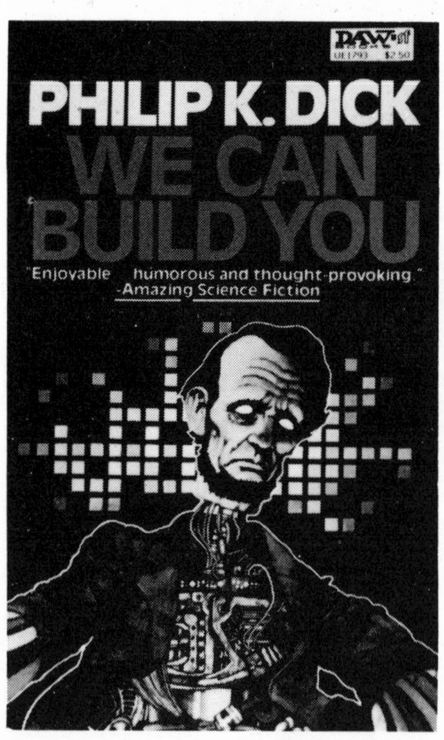

27

28

(29-30) Two shots from *Metropolis,* a 1927 German silent film. *Photos from the collection of William Kenly; published courtesy of Janus Films.*

CONSUMER INFORMATION:
Metropolis, a city in the year 2000, is beset by a working class dissatisfied with its living conditions. Industrialist Jon Frederson persuades a mad scientist named Rotwang to build a robot duplicate of the people's idealistic woman leader, Maria, intending to use the robot to incite the masses. Through Frankenstein-like apparatus, Rotwang takes his metallic mechanical woman to another dimension and makes a mirror image of her. In the *R.U.R.* tradition, violence ensues, but it backfires on Frederson, affirming a 1920s view that dire consequences await an individual who tampers with nature.

For science fiction film buffs, inexpensive photos of robots from the movies are available from Jerry Ohlinger's Movie Material Store, Inc., 120 West 3rd Street, New York, N.Y. 10012. Tel.: (212) 674-8474. Open every day, 1 to 8 P.M. Black and white photos cost $2 each. Color photos go for $4. To order by mail, write for Ohlinger's lists 44A (TV science fiction) and 44B (movie science fiction). Ordering instructions are on the lists. For posters of robot movies, request list #1, which covers films of the last ten years. "For older titles," says Ohlinger, "ask for specific titles. With these older films, posters are in and out very fast."

29

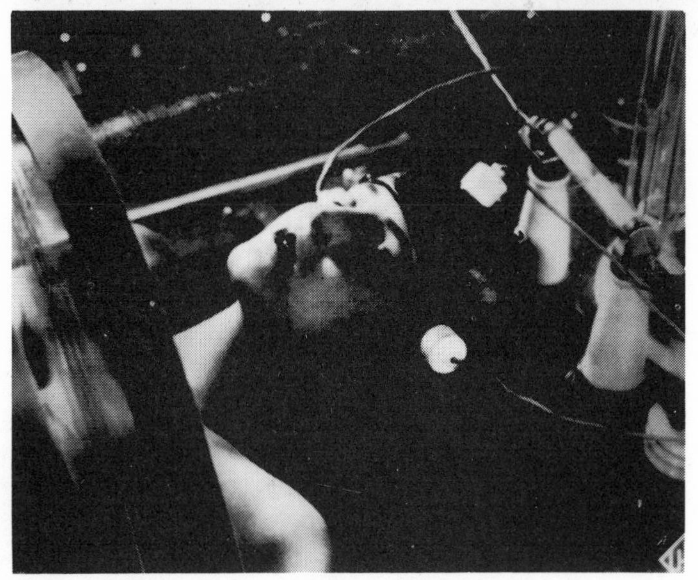

30

(31-33)
Early robot good guys: Tobor, from the film *Tobor the Great* **(31);** from *Forbidden Planet,* Robby the Robot **(32);** and Gort, from *The Day the Earth Stood Still* **(33).** Photo of Tobor courtesy of Republic Films. Photo of Robby courtesy of MGM. Poster photo of *The Day the Earth Stood Still* from author's collection.

31

33

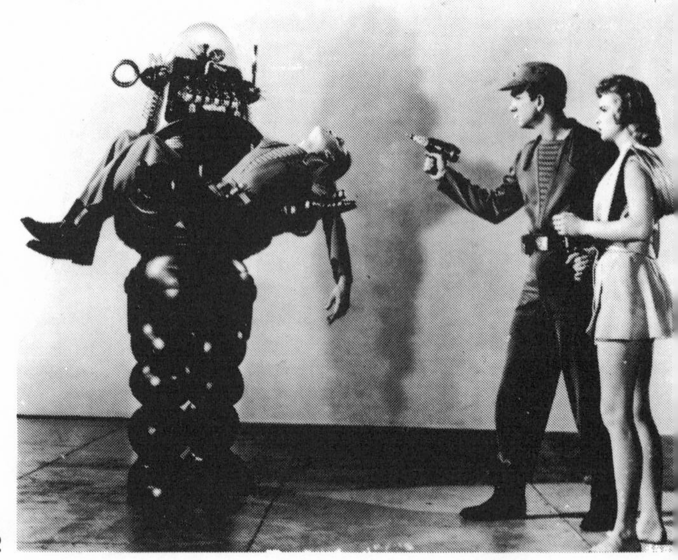

32

(34-37)
Robot bad guys:

(34) From the 3-D film *Gog* (United Artists, 1954), a robot of the same name went berserk at a space station in New Mexico.

(35) The Yul Brynner robot gunfighter in *Westworld* (MGM, 1974) was programmed to lose to tourists in a fantasy park. He malfunctioned, though, and turned trigger happy.

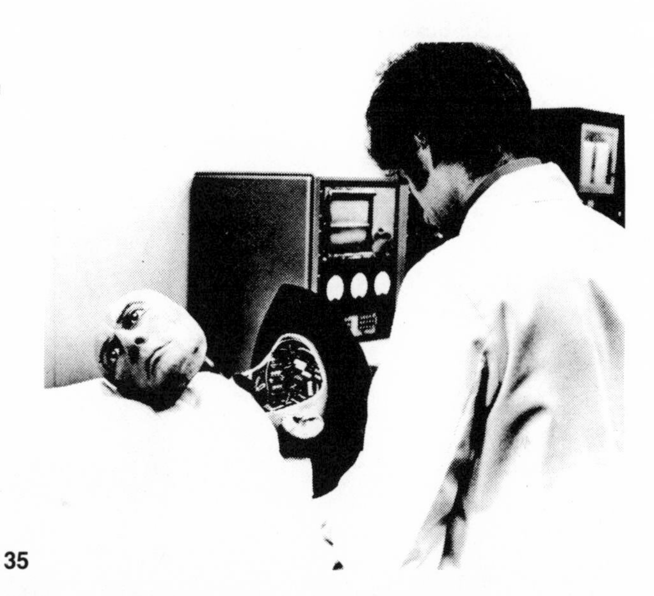

35

34

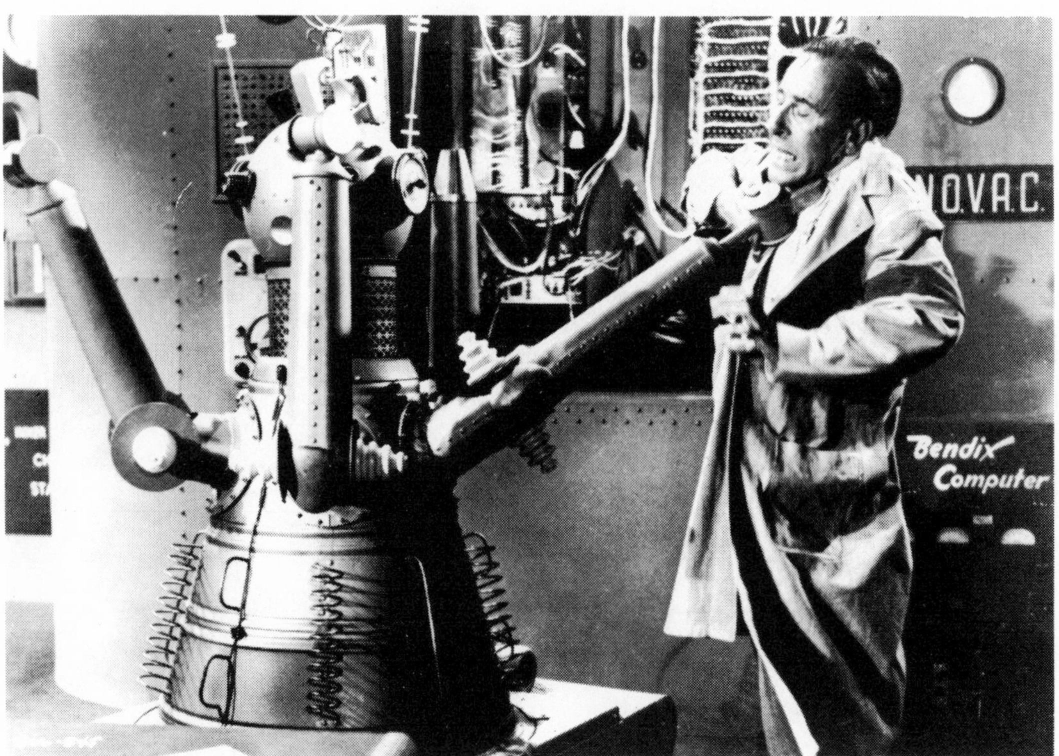

(36) The silver-faced Law Robots of *THX 1138* (Warner Brothers, 1971) maintained order in an oppressive society.

(37) The robot Box in *Logan's Run* (MGM, 1976) quick-froze would-be escapees from a futuristic domed city that was under the reign of a tyrannic computer. *Photo from* Gog *courtesy of United Artists. Photos from* Westworld *and* Logan's Run *courtesy of MGM. Photo from* THX 1138 *courtesy of Warner Brothers.*

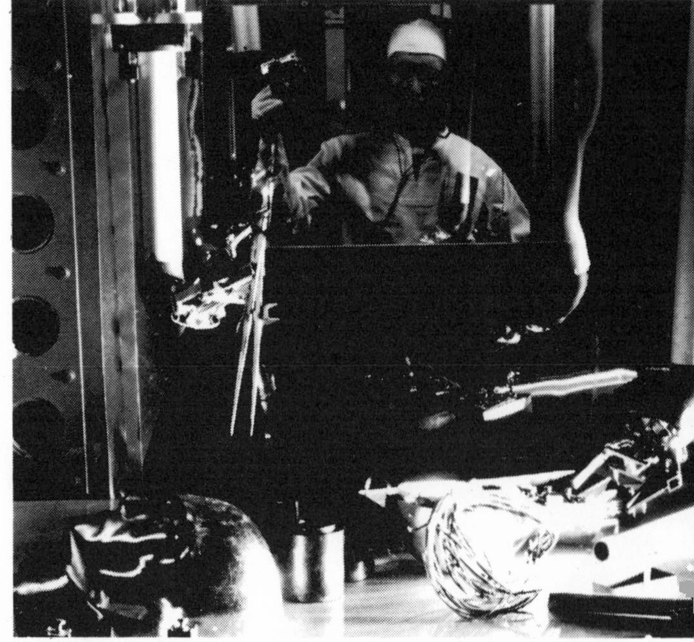

36

37

(38) A computer with designs on the Julie Christie character impregnated her in *Demon Seed* (MGM, 1976). *Photo courtesy of MGM.*

(39) In *Sleeper* (United Artists, 1973) health food store owner Miles Monroe, played by Woody Allen, awoke in a hospital 200 years after being cryogenically frozen. He escaped from the hospital in the guise of a domestic robot. *Photo courtesy of United Artists.*

(40) R2D2 and C-3PO, from *Star Wars* (Twentieth Century Fox, 1977). *Photo Courtesy of Lucasfilm Ltd., Copyright © Lucasfilm Ltd (LFL) 1977. All rights reserved.*

38

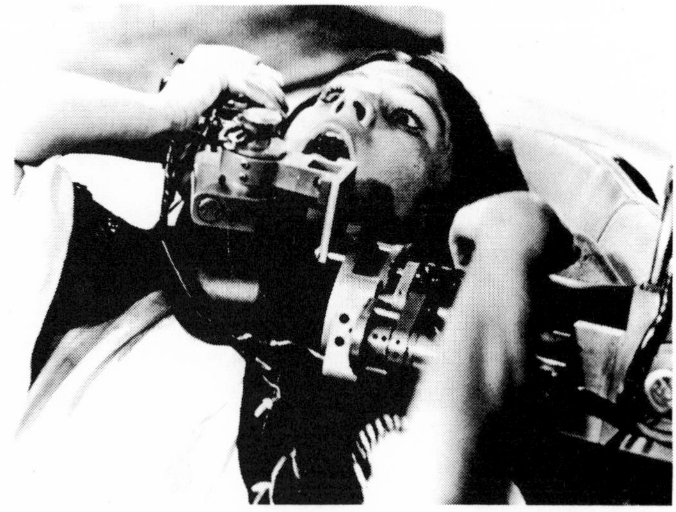

39

40

(41) From *Ballet Robotique,* a 1983 Oscar-nominated short. *Photo courtesy of Pyramid Film & Video.*

CONSUMER INFORMATION:
In this film without narration, giant manufacturing robots perform their computerized dances, pirouetting and sweeping majestically. The mathematical ballet of their fluid motions has been matched to romantic classical music, providing a unity between high technology and the arts. Selections include excerpts from Bizet's "Carmen Suite," Tchaikovsky's "Nutcracker Suite," Delibes' ballet "Sylvia" and, for the finale, the "1812 Overture," punctuated by showers of robot-directed welding sparks for cannon fire.

Ballet Robotique, an eight-minute color film, is sold in 16-mm or video cassette for $225. Rental: $55. Inquiries regarding sale or rental of *Ballet Robotique* should be directed to Pyramid Film & Video, Box 1048, Santa Monica, California 90406. Tel.: (213) 828-7577 or toll free (800) 421-2304.

41

(42) This cover for March 1949 *Astounding Science Fiction* was an idyllic look at robots. Copyright 1949 by Street & Smith Publications, Inc; renewed 1976 by The Condé Nast Publications, Inc.; used by permission of Davis Publications, Inc. *Photo by Phil Berger.*

42

(43-44)
Photographer Carl Flatow's photo **(43)** was originally commissioned for *World Magazine* and later used by *Omni Magazine* as the cover of its April 1983 robots issue. (See inset, **44**.)

CONSUMER INFORMATION:
"The concept I was trying to get across," says Flatow, "is of man creating machine in his own image. The idea was inspired by Michelangelo's *The Creation of Man."*

A limited quantity of full-color posters is available from Carl Flatow, 20 East 30th Street, New York, New York 10016. Color prints may also be ordered. Photo Copyright © Carl Flatow—1983. All rights reserved. May not be reproduced without permission.

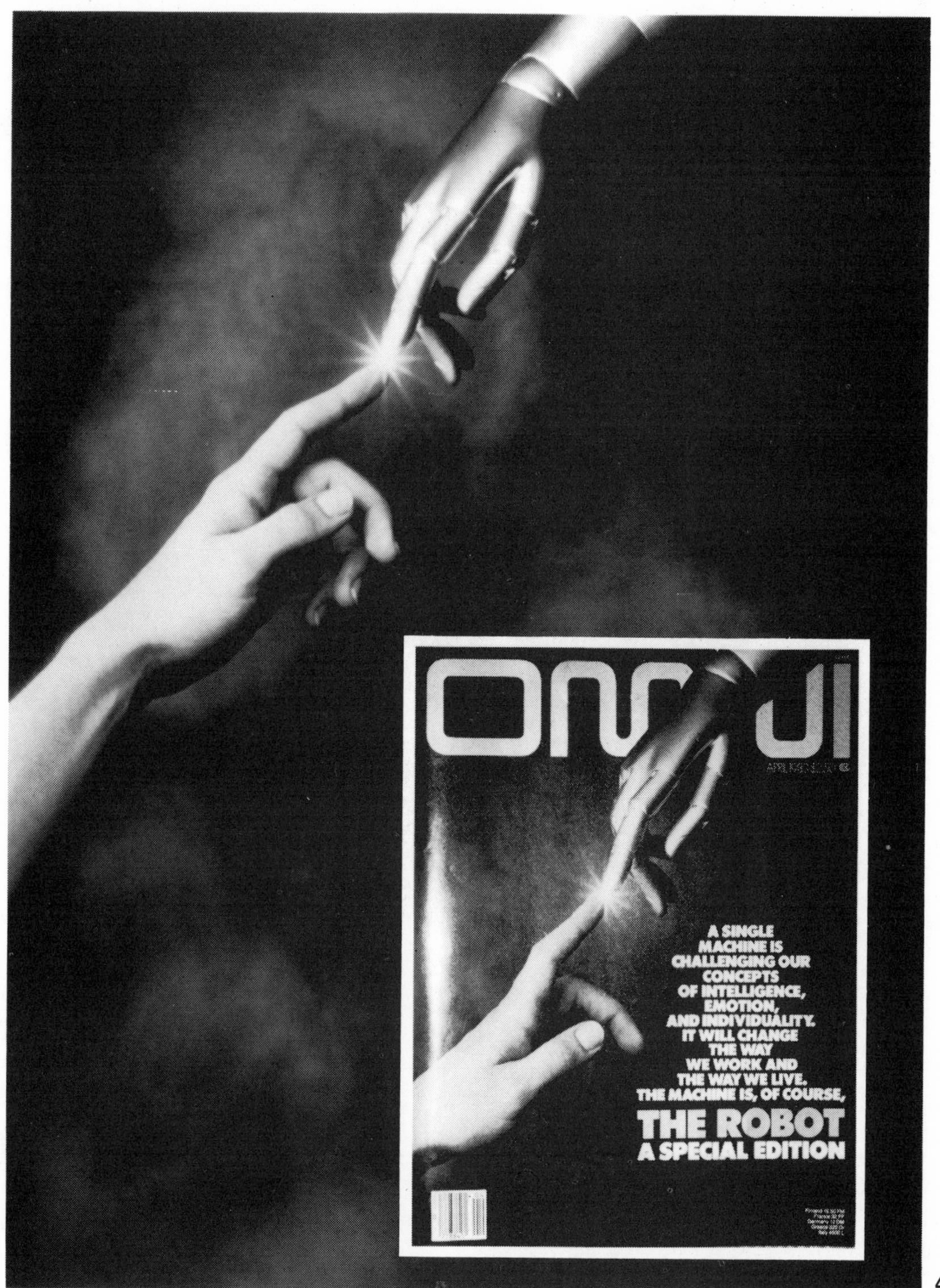
OMNI
A SINGLE
MACHINE IS
CHALLENGING OUR
CONCEPTS
OF INTELLIGENCE,
EMOTION,
AND INDIVIDUALITY.
IT WILL CHANGE
THE WAY
WE WORK AND
THE WAY WE LIVE.
THE MACHINE IS, OF COURSE,
THE ROBOT
A SPECIAL EDITION

43

FUN ROBOTS

Toy Robots, Show Robots, Personal/Home Robots

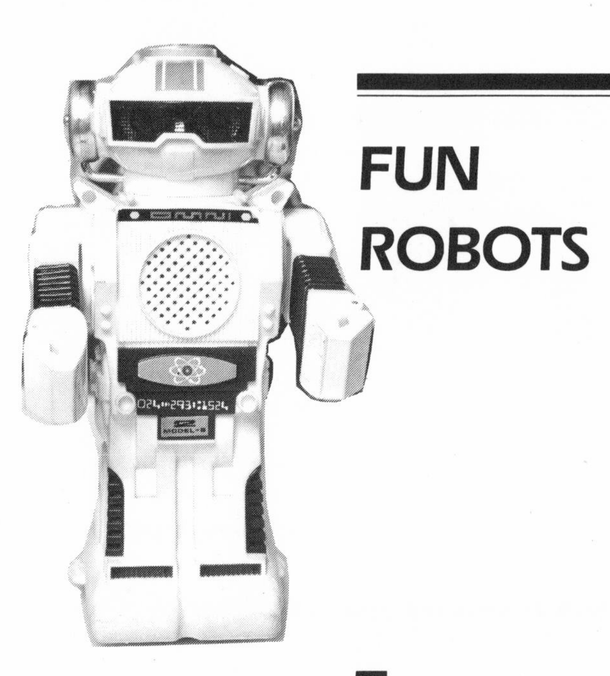

FUN ROBOTS

The robot that once needed an author's imagination or a Hollywood technician's special effects to come alive on the printed page or on a movie screen is no longer restricted to just fantasy.

Today the robot is, as they say, "live and in color"—the focus of a burgeoning industry dedicated to giving us, in one form or another, mechanical men that are diverting.

And diverting they are. Even in their most villainous movie roles, robots have always entertained us, much the way rogue wrestlers amuse us as, with cavalier disregard of the rules, they kick, gouge, and otherwise misuse opponents. The robot—as a force for evil or good—has a cartoon simplicity that makes it accessible.

That accessibility made the movie-going public embrace the *Star Wars* robot R2D2 and his somewhat pedantic colleague C-3PO. R2D2, the more popular of the pair, is often credited with the growing success of the various commercially made fun robots that are currently available—from those that are strictly toys to those marketed for personal, promotional, or educational use. Many manufacturers have seen their sales curves soar since 1977, the year *Star Wars* was first shown.

The irony is that R2D2 was not a truly mechanical figure but rather a shell for a small person who made it move. Of course that is a Hollywood tradition. In *The Day the Earth Stood Still,* the twelve-foot robot called Gort was inhabited by a seven-foot man named Lock Martin, who had once been a doorman at Grauman's (now Mann's) Chinese Theater in Hollywood.

Today's fun robots rely on simple wind-up mechanisms, or computers, or remote-control devices to move. Animated robots that entertain are not brand-new. Years ago, an isolated few were popular attractions. At the 1939 World's Fair, a speaking robot called Pedro the Voder worked the Bell Telephone exhibit. A spectator would state what he wanted to hear the robot say, and back would come the words, sounding a little unearthly but quite understandable.

Pedro the Voder was a device with "vocal cords" of vacuum tubes that reproduced vowels, consonants, and monosyllables. Pedro's operator used a keyboard and foot pedal to produce whole sentences of human speech from the mechanism. The robot reportedly could even do a Southern drawl or a Western twang.

At the 1940 World's Fair, Sparko and Elektro, a mechanical dog and man, appeared at the Westinghouse Exhibit. Sparko, it was reported, could bark, wag its tail, and even sit on its haunches and beg.

TOY ROBOTS

For years, the walking (and sometimes talking) robots that we saw were created for special occasions by big corporations like Westinghouse or Bell Telephone, or else they were built in the garages of hobbyists. The only robots that consistently moved off assembly lines were toy robots, which have a tradition that goes back to the nineteenth century toy automata. Some of the more recent toy robots are considered valuable.

Take one produced in the 1950s that is today considered a collector's item. Robert the Robot (45), manufactured by Ideal, is, when in mint condition, worth $800 these days.

His value is based on the innovation that the toy is said to have represented at the time it was marketed. Robert the Robot, you see, could speak. And while he was no sparkling conversationalist—"I'm Robert the Robot, mechanical man," was all he said—he was supposedly the first mass-produced toy robot able to utter words.

Nowadays, that is routine. Toy robots not only speak, they wink, they sparkle, they spin their parts and even emit smoke. There are toy robots that bounce off other objects, or that roll into a somersault. A toy known as Starry Robot (46) ejects yellow tokens from its mouth. Another, Saturn Robot (47), is able to shoot spring-action rockets off itself.

Robert, Starry, and Saturn are among the hundreds of toy robots for sale in a unique shop in downtown New York City. The Robotorium is, as its name implies, devoted to toy robots. In its narrow quarters (48), proprietor Debbie Huglin, a kinetic sculptor, keeps an ever-changing collection of old and new items. Her clientele, she says, is not (as one might expect) exclusively small fry. "In fact," says Huglin, "it's more like Wall Street types who keep The Robotorium going. And, of course, collectors. I have a large mailing list of collectors—thousands of collectors from both the U.S. and abroad. And sometimes I serve as a kind of middleman for them.

"Let's say Collector A writes in and tells me what robots he has and which ones he no longer wants to own. And then Collector B writes in and says what toy robot he's in the market for and what he's willing to pay. I match up the two parties and take 7½ percent of the selling price from each person.

"The collectors don't mind this because they prefer to pass on their toys to one another. They're very serious about collecting. Things like the condition of the box the toy came in, and how long the owner's had it—these are important matters for the collector. Like Robert the Robot: there are boxes for Robert that say 'patent pending.' Those boxes make Robert more valuable than later ones, which had the patent number on them. Many collectors I've known treat their toys as if they were made by da Vinci. In fact, I've known a couple of them who kept their little robots in a safe."

For the collector, or the merely curious, The Robotorium provides fresh stock on a regular basis, owner Huglin relying on her network of jobbers, suppliers, domestic and foreign manufacturers' representatives to locate old and new merchandise. "And what's happened," she says, "is because of the nature of this place any time there's a guy somewhere with a crate of toy robots he's looking to unload, I get a phone call to take it off his hands. Plus, I'll write companies here and overseas. I keep up, as any businessman will do. And through all my contacts I end up with a selection (49-57) that can't be equaled anywhere else. At any given time, I've got maybe three hundred to four hundred little robots on premises and thousands more in storage or in staggered orders."

Brand-name products and anonymous Hong Kong robots coexist on The Robotorium's shelves and counters. Some of them, like Robert, are the robotic version of an antique, which Huglin defines this way: "Any robot which has survived wars and kids and our general high-speed lifestyle for more than ten years is considered an 'antique.' "

New toy robots are personally inspected by Huglin before she adds them to her stock. That means she will actually disassemble each toy, partly because she enjoys tinkering ("I've been taking toys apart since I was four or five years old") and partly because it serves as her quality control. "I check each new type of robot inside and out," she says, "before I accept them. I want them to hold up, as I am selling them as more than toys."

For new collectors of robot toys, Huglin has this advice: "At the time you buy something, write down the name of the robot and then the manufacturer's name, as well as the date and store you bought it at. That will help to identify it later. Always leave the toy in the bubble pack or in the box in which it was shipped, and keep it where it can't get damaged or faded. After all, some day, your toy could be worth a lot of money. One other thing. Stay away from robots that are based on popular film characters, like the ones in *Star Wars*. You see, toy manufacturers have to pay so much for the right to the name of the movie robot that they can't afford to put design value in the toy."

Robots with a unique design concept—the Godaikin collection—have come to us from Japan (58-63). When fully assembled, the Godaikin are detailed space-age robots. Taken apart, they form space ships, vehicles, minirobots, or animals. There are ten Godaikin robots and they were first manufactured in Japan by The Bandai Group in 1973.

In late 1982, Bandai America, located in Allendale, New Jersey, began to distribute this series of robots throughout the United States. In describing the robots' warrior-like appearance, the manufacturer says: "Godaikin is the generic name for a group of strong and courageous robots. These robots are ready to battle any enemies, in order to preserve justice, peace, and love of the human race."

(Takara, another Japanese manufacturer, makes fierce battle robots that convert to cars. Takara calls its three-toy series Diakron Robot Cars of the Future (64).)

Not all robot toys are based on the science fiction concept of a humanized tin man. Take Armatron (65), a robot-like arm vaguely reminiscent of the scaled-down hardware that students and factory trainees use to familiarize themselves with robotic principles. But where, say, Colne's Armdroid (66) and Rhino's XR-1 (67) are priced from $995 and up—and are serious business—the mostly plastic Armatron, at its approximate retail price of $54, is strictly for fun.

Yet so sophisticated is Armatron that *Playboy* magazine, in reviewing it, said: "Tomy Toys may have labeled its Armatron game for 'Ages 8 & Up,' but we'll bet that the 'Up' end of the age curve is where the buying action will be. . . . The game pits you against an opponent—or a built-in timer—and the winner is the one who can maneuver liquid fuel canisters from one module to another without blowing up the world. . . ."

Says The Robotorium's Huglin: "Armatron is incredibly cheap, and incredibly high tech. It runs on a couple of D batteries (to power its electric motor) and is maneuvered by two joy sticks. It interfaces with your brain. To me it's the only robotic arm under $1,000 that's halfway decent."

Huglin suggests optional uses for Armatron. "You can," she says, "eat spaghetti with it, draw with it, smoke with it." *Playboy* has the last word, though: "Armatron is also a nifty way to drop an olive in a martini."

SHOW ROBOTS

A show robot is a souped-up toy—a remote-controlled mechanical figure that is usually life-sized and on the prowl for a good time.

Show robots—also known as entertainment robots, promotional robots, or communication robots—usually turn up at conventions and at shopping centers, their novel appearance and often witty remarks attracting crowds that end up being massaged in behalf of one product or another.

There is an aura about these robots that stems from the kind of species variance that made E.T. a universal hero. They are us—and they are not us. They are, at bottom, no threat and, because of that, can say the most outrageous things and not put off an audience.

To the archetypal good heart that these lumbering hulks are perceived to have, add the pleasure we take in the illusory sensation that the show robot is on his own, existing as freely in time and space as we do. If most of us know that somewhere nearby there is an operator—a professional actor probably—with a gizmo to propel his Denby (68) or Sico, or Six T Robot and give him a voice, we nonetheless suspend that awareness as we do in overlooking implausible twists in movies and plays, the better to savor the fun.

That fun, though, is a serious business to the handful of companies that make their profits from leasing or selling show robots. It is also a relatively new business, born of a technology that only lately has made these robots commercially feasible and of the need by businesses to draw attention to their products, or to enliven the sites from which the products are sold (69-70). The manufacturers' names do not echo tradition—no old-line robotic equivalents of Sears Roebuck or Ford Motors. In fact, most of those who preside over the field of show robots have stumbled into it.

Consider Gene Beley, president of Android Amusement Corporation in Irwindale, California. "Originally," says Beley, "I was a newspaper reporter (Ventura [Calif.] *Star-Free Press)* when I bought a Foosball table-top soccer game and installed it in a Ventura bar on a fifty-fifty basis back in 1967 when Foosball was more radical than robots today. I was just trying to make extra money on a newspaperman's salary.

"This began teaching me how to be in business for myself. I built a route of Foosball tables, then diversified with electronic games when I moved to Los Angeles. I began pioneering electronic game rooms in bowling centers in the early 1970s at a time when pinball had been illegal for thirty years in L.A. Then the State Supreme Court ruled them legal, and the business began taking off. I peaked at about 150 games and $225,000 a year. I then lost three of my accounts overnight when contracts weren't renewed, and I began reorganizing.

"I had seen a robot from Quasar Industries back East in New Jersey. I became their West Coast agent—like a Hollywood agent for a robot, I'd guess you'd say. It was very tough sledding prior to *Star Wars*. But after *Star Wars* debuted, it got easier. When Quasar chose to promote a domestic android, saying it would cook meals and clean the house and be in production in eighteen months, I split and began designing my own line."

Among Beley's creations was the remote-controlled DC-1, known as "The Drink Caddy" (71), built in 1980. DC-1 was a walking, talking globe-headed robot that dispensed cans of soda through its lower body and stored wine and champagne bottles in a rear compartment. It had a serving tray supported by arms that doubled as glass dispensers. Ice was kept in its transparent head.

DC-1 originally was sold through a Beverly Hills store for $6,000 in what Beley thinks may have been the first retail robot marketing effort in the world. His $3,000 wholesale price proved unprofitable, though, and he withdrew DC-1 from circulation.

Soon after, he was back with his more advanced DC-2 model (72). In place of the $17 water barrel that had served as DC-1's body, Beley installed a sleek fiberglass molded body with futuristic lines. As presently constituted, DC-2 is 4½ feet tall and weighs 194 pounds (crated 350 to 400 pounds).

Beley's other robots—Andrea and Adam Android—are radio-controlled, animated mannequin characters (73-74) that he rents. "I have turned down requests to sell them so far," he says, "because they are more fragile than DC-2 and I'm not anxious to be responsible for the servicing."

If Beley's odyssey from newsman to a robotics showman is unlikely, consider the transition made by David Colman. In 1965, Colman was a chorus skater in Shipstads and Johnson's Ice Follies. He rose to a role as featured comedy skater and, when the idea of a robot skater for the show was broached, Colman—who had been a mechanic with National Cash Register Company and had learned electronics in the Army's missile program—volunteered to build it.

What he wrought was a 7-foot 4-inch aluminum and plexiglas electronic robot called Commander Robot (75-76). Commander Robot could walk, talk, and skate. He was built from parts that included a 14-channel receiver, 7 motors, 50 pounds of batteries, clusters of servo switches, a tape recorder and amplifier, and assorted lights (ranging from dime-sized interior bulbs to a police car flasher winking atop a square-blocked head). The robot was built with a see-through body so audiences would immediately realize there was no human inside it.

So successful was Commander Robot's debut that Colman was elevated to prestigious backstage roles until, as shop foreman for the Ice Follies, he became responsible for the design and construction of lighting, sound sets, props, fireworks, and special effects (including robots). While he was with Shipstads and Johnson, as many as nine robots at a time were regular skaters in the shows.

These days, Colman is in business for himself with a line of robots that can be bought from The Robot Factory, his headquarters in Cascade, Colorado. In September 1982, one of Colman's mechanical men, Six T Robot, proved so alluring that on a trip to Dallas, Texas, he was stolen. Six T's disappearance inspired headline writers all over the country (e.g.: *Six T, phone home)* and prompted The Robot Factory to offer a reward for information leading to the arrest and conviction of the person or persons involved in the theft of Six T (77). Then, on October 20, 1982—more than a month after Six T had been taken—the director of security at a Dallas hotel received a mysterious phone call. "Look in your parking lot," an unidentified man said, before hanging up. Somewhat battered—and with his four-dollar bowler hat missing—Six T was found under a blanket where the caller said he would be.

Six T is one of many standard-model robots that The Robot Factory produces (78-81). All are remotely controlled and have effective operating ranges of up to 300 feet. Though Colman rents his robots, his profits are made through volume business—selling as many of his units as he can. Other companies use their corps of robots the way Avis and Hertz employ their autos: they make their money leasing their mechanical men.

At International Robotics Inc. a conscious decision was made early on to go after corporate accounts which, as IRI's president, Robert Doornick, concedes, was a calculated risk: "Because with our target market—the Fortune 500 companies—we were dealing with people educated to perceive such a robot as a quote, unquote 'gimmick.'

"My partner [Maris Ambats] and I had seen the robots that at the time—the mid-1970s—were popping in and out of shopping centers and trade shows, and we became interested in designing an exclusive and highly engineered robot—not to compete in the same market but to supply a sophisticated communication tool for Fortune 500 companies.

"We proceeded slowly toward that objective. About 1977, we built a prototype of Sico 1 [pronounced SEE-co] and built it entirely by hand and at great expense. Once we had built Sico (82), we realized we had no experience marketing robots. We had no history on how reliable our robots would be. And since we never wanted to embarrass the kind of companies we wanted as clients, we felt we needed a learning period.

"So we went to a market where the robot practically did not exist and where there would be no comparison. We went to Europe. We opened up an office in Paris to test Sico. [Doornick, a business major at New York University, was born in Paris]. And the day the robot landed, it started working for a big chain of department stores. And kept working."

Sico was upgraded and, by 1981, was ready for the corporate world on these shores. "In effect," says Doornick, "Sico was his own salesman, and was the best thing we had going in convincing the big companies to lease a robot. The way we did things, the robot would make his own calls. He'd step up to the reception desk and say, 'How do you do. I am Sico. I have an appointment to show myself.'

"And though, as I say, some of these companies thought of a robot as a gimmick, as soon as they had an opportunity to see the robot, their fears vanished. Because the psychological behavior between man and robot made it evident to them that there was value in a robot that could communicate with a customer.

"One other thing. When a technologically oriented company would use Sico at a business meeting, everybody would look at Sico and regard the technology behind him as having been developed by the company itself. And in the case of Sico, the technology is such that anybody would be proud to be connected with it. Our telemetry—that is, our control system—is different from any other robot's. Most robots have a remote control box that's related to the ones used for model planes—the two joy sticks and all. Ours is so miniature that it slips under the clothing of the actor who accompanies Sico. What that means is that when Sico works, the actor blends in with the crowd so that Sico appears to be walking and talking with no apparent equipment. Making him extremely impressive.

"In more detail, the actor has a hidden voice transmitter, with a microphone, under his tie or scarf or sleeve. And he speaks as a ventriloquist would, his voice a whisper. The robot's computer takes that whisper to a synthesizer to alter the voice quality—make it robotlike—and transmits it in its louder voice.

Among those companies that IRI says have used Sico are: TWA, Air France, Aero Mexico, Chrysler USA, AMC/Renault, Michelin, General Electric, Merrill Lynch, Pfizer, CBS, Coca Cola, Strohs, R.J. Reynolds, Scott Paper, Kodak, Walt Disney, American Express, and IBM.

International Robotics also will build custom-made robots that begin at $75,000 (83-85).

Like IRI, ShowAmerica Inc. of Elmhurst, Illinois leases its fleet of robots—85 percent of the company's business is in trade-show appearances. And while ShowAmerica has been in business since 1968, its president, Wil Anderson, came to it in the roundabout way that seems common to robot entrepreneurs.

"I would like to take credit," says Anderson, "for having created the concept of [show] robots, but in all honesty cannot do so. The first use of such an attraction, to the best of my knowledge, was by a part-time actor from L.A. In the late sixties, this fellow, Dave Cameron, linked up with an engineer from Kraft Model Aircraft factory in southern California and the two of them put together robots that were fiberglass motor boxes on wheels—radio-controlled, using model airplane controls.

"The first application that I'm aware of was for Falstaff Beer—I have a photo taken of a robot beer can working the sidewalk at Hollywood and Vine. The program did not go well, as the equipment was fraught with problems. The client quickly dropped the project.

"The next project that Dave sold was to a company introducing a new brand of cigarettes called Mavericks. At the time, I had a company called Promotional Enterprises, which handled publicity, promotions, and management of special events. I'd been doing that for several years. We were called in to manage a test program for Mavericks in San Francisco. Each day we would turn our four robots loose during factory-shift changes, at Fisherman's Wharf, at the airport, the financial district. The robots were moving around constantly, creating awareness for the product. The robots, built like large-scale Maverick cigarette packs, were able to vend samples to people.

"Following our management of the Maverick program, we undertook to represent Dave Cameron in the Midwest, but soon found that because of continuing equipment problems and his desire to sell the company to someone in the New Jersey area, equipment was not dependably available to us. We decided to go into the business on our own, and hired outside help to produce our early robots.

"Nowadays, all of our equipment is our own design (86), including the radio boards and all circuitry. We are in our seventh generation of electronics. We use no hobby equipment and no off-the-shelf material for our circuitry. Our units are state-of-the-art—all solid state, using FM radio systems. We have four voice channels so that we are able to shift channels should we encounter any local interference. Our units operate very efficiently, and a single battery provides sufficient operating time to run a ten- to twelve-hour show day.

"We retained the motor-box-on-wheels modular concept and thereby are able to place various corporate or product shells on top of the motor box—we call it a Robacon (87)—and use the modular base as a propulsion system. As a result, we can place a replica of a client's product—be it a can, box, bottle, cartoon figure, etc.—on top of the motor box (88-91). Our materials travel as excess baggage in fitted cases, and we have flown everywhere. To date, ShowAmerica robots have appeared in twenty-three countries. Our volume of business is at the point today that we have at least four robots appearing somewhere in the world every day of the year on behalf of a client. Much like the old cliché of the British Empire, we are at the point, worldwide, where the sun never sets on a ShowAmerica robot—and that includes our latest robot, Peeper, which is a humanoid design with a periscoping neck (92)."

When entrepreneurs like ShowAmerica's Anderson take their robots to the public, strange things sometimes happen. "One robot," says Anderson, "upon arriving in Saudi Arabia, encountered a superstitious customs agent who would allow us to keep the robot body, but would not admit the head of the robot, which is removed for packing purposes for shipment. Fortunately, we were on a U.S. Department of Commerce sponsored tour, working on behalf of Westinghouse Electric. Our robot operator contacted the American Embassy and, after two days of negotiation with the Saudi government, the head of the robot was brought into the country under diplomatic immunity.

"On another occasion in Salt Lake City, a young man in a shopping center, named Tyler, became very excited upon meeting the robot and started stuttering rather badly—to such an extent that he had difficulty talking with the robot. Luckily, we had a good perceptive operator who, through the robot, suggested that if Tyler would calm down and take hold of the robot's hand the robot would give him a burst of kinetic energy that would enable him to talk without stuttering. By this time, a crowd had gathered around to witness this, including—unknown to us—Tyler's father. Tyler followed the robot's instructions and there was enough of a psychological factor working in our behalf that, sure enough, Tyler was able to talk without stuttering. Afterward, Tyler's father came over and thanked the operator for his graceful handling of the situation and remarked on the unusual psychological effect the robot had had on Tyler, expressing the desire that some of this would carry over to help the youngster, who had had a stuttering problem all of his life."

Occasionally incidents occur with these robots that provoke newspaper headlines. It happened when the sons of Android Amusement's Beley operated a DC-2 from a concealed position on a public street in Beverly Hills. When police officers spotted the robot and couldn't locate its operators, the DC-2 was hauled off to jail. An apologetic phone call from Beley's wife, Jill, and a $40 towing charge obtained the robot's freedom.

By contrast, International Robotics' Sico is trained to move through public places on his way to a job. And that includes stepping up to an airline counter and showing his Diners Club card (3852 208502 0018), which is made out to ROBOT SICO. "He flies first class," says IRI's Doornick. "Not because he's a snob. FAA regulations state that any equipment has to fly forward of any other passenger. So Sico always gets first row, first class. There's not one airport where he doesn't meet thousands of people. It's good promotion."

It is not unusual, according to Doornick, for robots like Sico to trigger extreme reactions in people: "I brought Sico to a hospital in England once. And I remember we rounded the corner on one floor and came face to face with a nurse from the West Indies. She ran away, screaming 'Voodoo, Voodoo!' and quit her job that day. On the other hand, the robot sometimes jokingly invites a woman up to his room, and there have been times when women actually have shown up."

PERSONAL/HOME ROBOTS

Toy robots appeal to us for their ability to produce mechanical effects. We respond to their quirky charm, their somewhat comic mimicry of our own movements.

Show robots entice us with the ingenuity of their performance, the technologically virtuoso moments they provide.

But neither mechanical type taps the fantasy that we humans have attached to robots, that mythological notion that some day mechanical help would labor on our behalf, performing household drudgeries that would free us for more exhilarating activities.

For years, the idea that such a creature could be fashioned existed as a Buck Rogers vision of the future, a speculative picture that seemed more charming than actually possible.

But by the 1970s, the computer revolution had brought the idea of personal robots within the range of manufacturers' technological capabilities. The problem at first was cost: it simply was too expensive to produce home automatons. But with the plummeting price of silicon chips, suddenly a robot to do the master's bidding became affordable.

In 1983, these personal robots came on to the market, triggering reams of copy in the print media (93).

Here Come the Robots, intoned *Time* magazine, which wrote: "Rolling across the floor on big black wheels, it embodies one of man's most enduring dreams: the personal robot . . ." That hallelujah tone was common to the prose heralding personal robots with names like Hero-1, Genus, and Topo. To a degree, some manufacturers added to the hype. Androbot, Inc. of Sunnyvale, California, in its brochure, with a cover showing a robot gliding across a desert (94), announced:

> It is time.
>
> The Year 1 A.B. opens a new era in technological history—a dream that has engaged man's imagination for centuries.
>
> The microprocessor—a device that has made the personal computer (with the capability that only five years ago required a mainframe office computer) now makes possible the personal robot, and a profound change in our lives is about to take place . . .

The excitement of bringing forth a new and intriguing product could excuse a company's sounding a bit goggle-eyed. But the effect of the many trumpeted acclamations was to distort what was really happening. For the impression created was that the product—this first wave of home robots—was the masterwork home helpmate futurists had envisioned. In fact, it was, and is, but a modest beginning.

"If you expect it [the personal robot] to wash windows or clean dirty dishes, you're going to be disappointed," Douglas Bonham of the Heath Co. (maker of Hero-1) told interviewers.

The sobering truth is that the new robots are not domestic factotums. In fact, Heath's promotional literature presents its Hero-1 from a less glamorous perspective—as an educational tool (95):

> There's no escaping it. The worldwide direction for industry today is robotics. All across North America, Europe, and the Far East, industry is turning to robots to perform a multitude of manufacturing activities in order to remain competitive . . . We are at the start of a robotics revolution and there is a tremendous shortfall in educated, trained personnel to design, specify, sell, program, operate, install, and service these robots.
>
> The Heath robot makes it possible for nearly anyone to obtain a comprehensive background in robotics.

Hero-1 as a hands-on experience in robotics was what Heath was selling. And though personal robots like Hero-1 do walk, talk, sense their surroundings and, when equipped with grippers, pick up objects, early reports indicate that they are not yet entirely reliable.

The New York Times science writer William J. Broad built a Hero-1 from scratch, assembling the 1,200 parts of the $1499.85 Heathkit, including 150 semiconductor chips, a computer brain, and eight motors. "Hero," wrote Broad, "taught me a lot about his strengths and weaknesses, and those of his android brethren. He could easily function as an intruder alarm or say funny things and twirl about on his tricycle wheels. But he had difficulty in dealing with his surroundings—with things a human finds quite simple."

Broad discovered that when Hero navigated a room in an attempt to pick up a can of Pepsi, his accumulated errors "would often leave him clutching at thin air."

Daniel J. Ruby of *Popular Science* experienced similar problems. "For fifteen minutes I sent Hero on a herky-jerky chase around the room before I finally got it to pick up a film canister . . . placed on the floor."

On a visit to Androbot, Inc. for *Playboy,* David Owen reported: "Soon we heard a whirring noise and turned around just in time to see an honest-to-God three-foot-tall beige robot get stuck in the shag carpeting and make a sound like a car spinning its wheels on a sheet of ice."

These items—gleaned from a large body of print materials—are not meant to damn the current crop of personal robots, but *are* intended to caution prospective buyers that glitches can and do turn up in what is, after all, an evolving industry. Robots at this stage are very much "in process."

Take the RB5X (96), produced by the RB Robot Corporation. By mid-year 1983, RB5X had a refinement it had not possessed in the original model that was offered at the first of the year. The manufacturer reported: "The [new] RB arm has five axes of motion and folds completely inside the robot when not in use. It can be controlled directly by software or programmed using a separate teaching pendant, which attaches by cable to the robot and then detaches once the arm learns a particular task. In addition to lifting and carrying up to twelve ounces, the arm can also be programmed to reach around and turn the robot off." (Later, RB Robot Corporation announced that an optional vacuum attachment [additional cost $595] would be available before the end of 1983.)

For prospective buyers of intelligent robots, the variations from brand to brand and the wide range of prices can be confusing. A personal robot ought not to be an impulse purchase. Study the prospectuses of the manufacturers and take advantage of dealer demonstrations when offered.

The State-of-the-Art Robot Catalog offers a simplified guide to the personal robot market. However, though the guide gives consumers a solid basis for comparison, they would do well to consult manufacturers regarding late-breaking price and specification changes.

Manufacturers are listed alphabetically.

ANDROBOT, INC.

Chairman of the Board of Androbot, Inc. is Nolan Bushnell, who founded Atari and the Pizza Time Theater restaurant chain. His arrival on the robot scene suggests commerical magic may be in the offing (97).

Androbot, Inc.'s original robots (Topo and B.O.B.) were designed, according to the manufacturer, to provide entertainment and education. As Androbot's promotional brochure says: "A personal robot—as we see it at Androbot—should be play-oriented, rather than work-oriented. Not an appliance, but a friend."

In its ABS plastic body and steel base, Topo (price: $795) stands 36½ inches tall and weighs 33 pounds. It is a mobile extension of the home computer, which actually serves as its brains and memory. Designed to interface with most popular personal computers via a remote IR (infrared) communications link, Topo's current capabilities for text-to-speech talking and movement make it a tool for parents and teachers to encourage children to sit at the keyboard and learn to program.

Two methods of control enable the user to direct Topo's actions. Teach Mode provides for either joy stick or keyboard entered commands to instruct the robot in what to do or say; any sequence of commands thus ordered may be imprinted simultaneously on disks, to be retrieved later at the user's option.

In addition, Androbot also supplies a software package based on a standard version of FORTH, which enables the user to write original programs for Topo. Challenging interactive educational games—which have Topo posing questions and rewarding answers verbally or by performing a playful routine—are one major category of programmable software possible at this time.

Similar to another Androbot, B.O.B. (Brains on Board), Topo is equipped with an internal "bus" system which will allow for future upgrades via newly developed software and electronics. Aiming at educational markets, Androbot began selling Topo in April 1983.

Topo's more sophisticated cousin, B.O.B., was due to go on the market in late 1983. B.O.B. (price: $2,995) stands 36½ inches tall and weighs 42 pounds, and is constructed from ABS plastic, with steel used in its base. B.O.B's on-board "native intelligence" derives from Intel 8086 microprocessors combined with 3-megabytes of memory capacity. B.O.B. will navigate a living space and talk in a human-like voice, randomly choosing from over one hundred stored words and phrases. Infrared sensors attract B.O.B. to humans, whom it may follow at will. In the process, it will avoid inanimate objects in its path, through its ultrasonic sensing devices. One additional feature of note: B.O.B. can retrieve a beer or soft drink from an optional Andro-Fridge, and bring it to wherever its master may be waiting.

Equipped with an exclusive Androbus system, B.O.B. has the potential for expansion through add-ons to its existing electronic brain and through user created and commercially available software. At the core of the Androbus are ten slots, some of which house the electronic components that comprise B.O.B's current operating system. The remaining slots will be able to accommodate a range of additional cartridges and plug-in boards, for memory and CPU [central processing unit] expansion, for example, as well as for specific functions (i.e., voice recognition or text-to-speech modules). B.O.B's Androbot Control Language (ACL) provides a high level programming capability for the serious enthusiast.

According to the manufacturer: "In the future, B.O.B. has the potential for being not only a charming companion, but also an indispensable multitasking computer on wheels. B.O.B. may sound a wake-up call, and coax late sleepers out of bed in the morning; watch over the house; take phone messages and respond to emergencies while the family is away at work or school; and entertain them in the evening after a long day with a recitation from Byron, Keats, or Shelley. Its capabilities are limitless, depending only on his owner's programming prowess, enhanced by Androbot-provided hardware accessories and third-party software."

Androbot is currently marketing an AndroWagon (price: $95), which easily attaches to B.O.B's midsection, and enables it to transport books, beverages, or the baby's diapers from room to room. Constructed of the same durable plastic material used for B.O.B. and Topo's body shells, the AndroWagon will be an available accessory for both Androbots. Other software and accessories that were under development at midyear 1983 include an AndroSentry home security/alarm package, and AndroFridge.

In addition to B.O.B. and Topo, Androbot was scheduled to introduce, before the end of 1983, a small-sized robot targeted for the 6 to 14 age group and priced at under $300. Called F.R.E.D. (Friendly Robot Educational Device), the robot is a compact and sturdy mobile extension of the home computer, but also can be operated independently by a remote infrared controller, thus broadening its use to consumers who don't own a computer.

When placed on a tabletop or other flat surface, F.R.E.D. can be programmed to perform any series of movements without tipping off, since its mechanical sensors automatically tell it to avoid dangerous edges. Place a sheet of paper or white poster board under F.R.E.D. and, using its drawing-pen attachment, this Androbot can execute precise renditions of complex geometric shapes designed on the computer screen.

Packaged with F.R.E.D. will be a mini AndroWagon, enabling him to transport small items from room to room. Planned options, including an accessory arm, are currently under development, and will provide F.R.E.D. with additional abilities and function. F.R.E.D.'s design allows for expansion via future software, such as a voice synthesizer for speech programmable by the user.

HEATH COMPANY

Heath Company produces Hero-1—an acronym for Heath Educational Robot. (price: $1,499.85 in kit form; $2,499.95 fully assembled). Also available: a kit form Hero-1 without arm and speech synthesizer options. Cost: $999.95. Hero-1 (98) is 20 inches tall, weighs 39 pounds and, with its rotating turret-like head, bears a resemblance to *Star Wars'* R2D2.

The robot has an on-board 6808 microprocessor that guides it through its maneuvers and activates sensors that can detect sound, light, and motion. Ultrasonic sensors are able to determine the range of an object up to eight feet away. With an optional ET-18-2 Phoneme Speech Synthesizer ($149.95), the robot can simulate human speech, with four levels of inflection. It moves on a three-wheel drive system and can turn in a twelve-inch radius.

Hero-1 is programmed in three different ways: through the keyboard mounted on the robot's head, with its hand-held remote-control teaching pendant, or through its serial cassette port (using programming previously stored on a conventional audio cassette tape recorder). The computer can store programs with more than one thousand individual steps.

The robot can be programmed to detect intruders in its range and warn them away verbally. And Hero-1 can remain on guard for extended periods of time, using its power-conserving "sleep" mode.

Hero-1 comes with an optional ET-18-1 arm and gripper mechanism ($399.95) which lets the user program the robot to pick up small objects. The arm extends, retracts, and turns as it performs its mechanical tasks. The gripper can hold up to a pound when the arm is fully retracted and horizontal.

Hero-1 has on-board rechargeable batteries and an external battery charger.

Designed as a robotics and industrial electronics trainer, Hero-1 permits easy access to all interior boards and components, and there is an optional 1,200-page self-instruction text, covering robot fundamentals ($99.95).

RB ROBOT CORPORATION

RB Robot Corporation is the maker of the RB5X Intelligent Robot (price: $1,495). RB5X (99) stands 23 inches high, weighs 24 pounds and has an aluminum chassis and polycarbonate dome. It moves on two 4-inch diameter synthetic rubber drives and two 2-inch diameter casters, which allow turning either on center or off center.

RB5X may be programmed using any computer with an RS-232 communications interface. Once connected, the computer serves as a "dumb terminal" for altering the robot's programs or studying its memory. If then disconnected from the computer, RB5X operates completely on its own.

The robot is equipped with standard sonar and tactile sensors that enable it to detect and react to objects in its path. Its instinct is to keep moving. Encountering an object for the first time, it chooses from a table of random responses and stores in its memory the response that allowed it to continue in motion.

As its experience grows, the RB develops ranking, or levels of confidence, in each of its possible responses. Eventually, the RB creates a range of appropriate learned responses to all of the circumstances it encounters in a room. When its memory is cleared, the RB5X may establish an entirely new and equally effective set of learned responses to the same environment.

A special circuit in the RB enables the robot to recognize that its batteries are low and to seek out its charger. It then automatically charges its batteries, detaches itself from the charger, and resumes its activity.

RB5X is equipped with pulsating lights that can be programmed to correspond to whatever mechanical or electronic events the user chooses. This feature, along with RB's standard horn, can alert the owner to special circumstances, or can be used simply for extra interest.

In addition to the recently developed optional arm, RB Robot Corporation offers voice recognition and speech synthesis options. The voice synthesis option is based on a standard synthesis chip which provides a speaking vocabulary limited only by the size of the robot's memory. A speaker inside the robot and a programmable sound generator also enable the robot to reproduce any sound, such as music, whistles, or sirens.

To increase the robot's learning capacity, RB owners may add an optional 16K bytes ($125) to the standard 8K bytes.

ROBOTICS INTERNATIONAL CORPORATION

Robotics International Corporation has a home robot called Genus (GEE-nus) that was to have gone on sale by mid-1983, but did not. Plans now call for Genus to debut early in 1984. According to Joe Collins, director of special projects, "We're building our own microprocessor and that's the big reason Genus has been delayed. The microprocessor will have 200K capacity."

Made of high-impact plastic and Fiberglas, Genus (100) is 4½ feet tall, weighs 120 pounds, and its large scale is carried through in pricing—$10,000 for a model that is expected to include a vacuum unit and a security package. A more budget-minded Genus may be offered too.

The robot is mechanized by two motor-driven 7-inch rubber wheels with fore and aft stability provided by casters. On-board microprocessing allows Genus to learn its surroundings without human assistance. Its ultrasonic obstacle avoidance system prevents Genus from bumping into walls, furniture, or pets.

When its batteries are low, Genus seeks out and connects to its own 110v outlet. Two 12-v sealed lead acid batteries provide an average time between charges of four hours.

Information can be called up on Genus's built-in CRT, and the robot can be commanded—depending on optional packages—to shake hands, speak, provide security against intrusion, fire, gas leak, and water.

The vacuuming unit hangs on the front of the robot and features a beater brush housing that moves forward and backward on a bellow arrangement. With its map of the home, or a user program, the robot works from room to room, vacuuming without need of human monitoring.

Advanced users will be able to write programs and create video games or educational quizzes that make the robot an active participant.

TECHNICAL MICRO SYSTEMS, INC.

A related item in the personal robot market is the curious computer-carrying "turtle" called Itsabox (101) that can manipulate objects around a tabletop. Anticipated price: kit version under $500; assembled $600.

Under development by Technical Micro Systems of Ann Arbor, Michigan, Itsabox measures approximately 6″×8″×4″ high and weighs about 3 pounds. It is completely self-contained, requiring no connecting cables or remote computer.

The on-board CMOS computer can be programmed in BASIC, FORTH, or other high-level language. An RS-232 interface is provided so that Itsabox can communicate with any standard computer or terminal.

Itsabox can move about freely on a table, exploring as it goes with its vision and touch sensors for objects, walls, or table edges. With its hand, it can pick up small objects and move them about, rearranging them according to the user's instructions.

Itsabox's hand has two horizontally opposed fingers. The fingers can be opened and closed by a stepper motor, and a pair of sensing switches are included to recognize when the fingers have closed on an object. The finger switches also double as wall sensors when moving and turning. Once Itsabox has grasped an object, the closed hand will lift the object off the table. Itsabox can grip objects up to two inches in diameter and lift objects up to four ounces.

Also included is a loudspeaker for making busy noises while the robot is on.

According to Lee A. Hart, of Technical Micro Systems, Inc., the marketing of Itsabox came about inadvertently: "The response we're getting on Itsabox has caught us by surprise. We had developed the robot as a demonstration tool for our BASYS family of CMOS computers and the associate Proteus development systems. When we included the Itsabox flyer with our Proteus press release to *Computers and Electronics* magazine, they included it in their January [1983] robot article as if it were in production when it wasn't. Needless to say, this produced a tremendous number of inquiries that we are still trying to dig out from under. Since then, we've been working to develop Itsabox into an actual product. It keeps evolving almost daily."

The last word on home robots may very well be Companion, which, as conceived by Robert Doornick of International Robotics, Inc., is a luxury robot priced at $150,000 to $250,000, depending on the options chosen.

Companion (102) is seen as a multipurpose robot: a computer; a home entertainment center (with color video monitor and TV set that plays video games as well as dials in conventional stations, stereo system, video tape recorder, color video camera for taking home movies, wireless telephone); a complete security system (for detection of intrusion, fire, poison gas, and heat).

"Not only will Companion be able to do your accounting," says Doornick, "and store general information like recipes . . . not only will he be able to give you a night at the movies or the latest record album . . . he'll do all that and, when you press the button for security mode, he'll guard your home, phoning you after you leave if he detects an intruder for a determinate period of time or fire, heat, or gas. He'll instantly dial the phone number you've punched in before leaving and automatically will deliver a message: 'Security break, please alert so-n-so.' And he'll hang up and he'll keep dialing 'til someone has come home and altered him. He'll keep dialing if the line is busy. We're going to make him a stubborn robot.

"At the same time, Companion will have turned on his video tape recorder and color video camera and sound recording devices and will scan in the direction of the occurrence and record visually. You can play it back later and understand how things started.

"In addition, Companion will be programmed to go from room to room without the intervention of the owner . . . in a fully automated manner, making sure to stop if a cat is in his path. He'll have a docking module, which will be placed on the wall of one of the rooms to recharge on his own."

Companion will be custom-built, and available through Sperry & Hutchinson who, in addition to its business in S&H Green Stamps, has a catalog of merchandise that corporations and companies use in incentive programs for their employees. "This catalog," says S&H's Larry Smith, vice-president of marketing, "is used as a way to increase productivity, or decrease absenteeism. To get a shot at Companion, you'd have to be part of a program sponsored by a company using S&H's incentive program. The merchandise in the catalog we think of as being on par with Bloomingdale's, Neiman-Marcus, or Horchow's. In other words, special. And Companion certainly will fit in that category."

The Sperry & Hutchinson catalog, whose offerings include Companion, was to become available in late 1983.

CATALOG AND CONSUMER INFORMATION

(45-48)
Robert the Robot **(45),** an early talking robot; Starry Robot **(46),** which ejects tokens through its mouth; and Saturn Robot **(47),** which propels spring-action rockets from its body are among the items either on display or for sale at The Robotorium, a store devoted exclusively to toy robots. The store's proprietor, Debbie Huglin, is pictured on the premises in her special-occasion gallactic headwear **(48).** *Photos by Maggie Steber, courtesy of Robotorium, Inc.*

CONSUMER INFORMATION:
Collectors or other interested parties may obtain The Robotorium's free catalog by writing to Debbie Huglin, The Robotorium, 252 Mott Street, New York, N.Y. 10011. Tel.: (212) 966-6881. Starry Robot retails for $18 and Saturn Robot for $20 at The Robotorium. Those desiring to buy, or sell, collectors' items like Robert, should contact Ms. Huglin in writing. She puts buyers and sellers in contact with one another and takes 7½ percent of the price of sold items from each party.

45

46

47

48

(49-57)
From the stock of The Robotorium: *Photos by Maggie Steber, courtesy of Robotorium, Inc.*

(49) "Robot 7" is an all-tin robot that moves while erect or on its stomach. Made in Taiwan, 1968. Retails for $5.

(50) "Tommy Atomic" walks, talks ("I am the atomic-powered robot. Please give my best wishes to everybody"), and emits a siren-like beeping. Made of quality plastic, Tommy Atomic spins away when it contacts an object. Lights glow in his eyes. Made in Hong Kong, 1981. Retails for $25.

(51) "Sparking Robot" is made in Hong Kong and retails for $2. Sparks are emitted from its mouth.

49

50

51

(52) "Acrobot" is a tiny wind-up robot that walks, falls, and returns to his feet. Huglin says: "Acrobots are fragile and hard to find on the market. I can see them in ten to twenty years being worth $1,000 apiece." Made by Tomy Corporation of Carson, California, circa 1978. No longer manufactured. Retails for $7; $15 in its bubble pack.

(53) "Sumo Robots" are made by Tomy-Japan, but not distributed in the United States. The Robotorium has them on display, but not for sale at this time because of copyright regulations.

(54) "Gyrobot" is a gyroscopic-powered toy that spins, walks, and stands upside down. Made by Tomy Corp. of Carson, California. Retails for $4. Appeared on market, 1983.

(55) "Machine Gun Robot" has a heating element that turns a drop of 3-in-1 Oil into smoke that comes out of the toy's chest. Made in Hong Kong, 1982. Retails for $18.

52

53

54

55

(56) R2D2 is no longer in production. The *Star Wars* hero was manufactured by Kenner Corp. about 1978. R2D2 is not mechanical. Retails for $18.

(57) C-3PO is not mechanical. Retails for $18. *Photos by Maggie Steber, courtesy of Robotorium, Inc.*

56

57

(58-63)
The Godaikin, a series of ten vividly multicolored warrior robots, seen here in closeups of Gardian **(58)**, Daltania **(59)**, and Golion **(60)** and in poster views of Sun Vulcan **(61)**, Voltes V **(62)**, and Combatra **(63)** that show how the robots disassemble into other objects, ranging from lions to military vehicles to missile-ejecting spaceships. *Photos by Phil Berger, courtesy of Bandai America, Inc., 6 Earl Court, Allendale, N.J. 07401.*

CONSUMER INFORMATION:
All Godaikin toys are made of die-cast metal and high-impact plastic. Some assembly required. Recommended for ages 12 years and older. The Godaikin robots are manufactured in Japan by The Bandai Group and distributed in the United States by Bandai America. The toys vary in price. Suggested retail prices range from $45 to $85. Voltes V ($85), when disconnected, becomes assorted military machines; Sun Vulcan ($45) converts to a missile-ejecting spaceship and a crane-toting bulldozer; Combatra ($85) becomes several military machines.

58

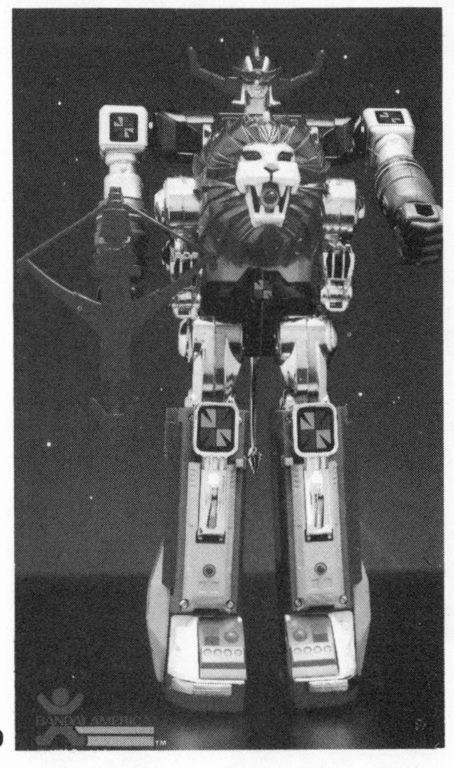
59

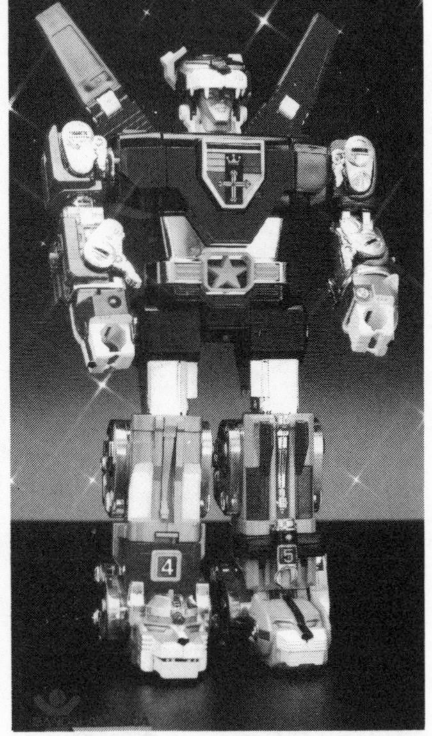

60

SUN VULCAN™
GoDaiKin™
An Innovative Series of Super Robots
Item #77065
• SUN VULCAN™, an amazing 9" colorful robot, is a fascinating example of modern machinery.
• SUN VULCAN's torso, with a few touches to some buttons and flaps, becomes COZMO VULCAN, a missile ejecting spaceship.
• SUN VULCAN's legs, when snapped together, form BULL VULCAN, a crane toting bulldozer.
• SUN VULCAN™, which comes with a wide array of accessories, is made of durable die cast metal and high impact plastic.
• No batteries required.
• Some assembly required.
• Recommended for ages 12 years and older.
• Ctn. Pk. Pcs. 6/Ctn. Wgt. Lbs. 16/ Cu. Ft. 3.2
© 1982 Bandai America

61

VOLTES V™
GoDaiKin™
An Innovative Series of Super Robots
Item #77052
• A pillar of strength and energy. VOLTES V™ is a fascinating robot which stands 13" tall.
• VOLTES V™, formed when five distinctive vehicular robots are united, is constructed of durable die cast metal and high impact plastic.
• VOLTES V™, when disconnected, becomes five assorted military machines: CREWZER I, BOMBER II, PANZER III, FRIGATE IV and LANDER V.
• VOLTES V™, which can be transformed into a space age tank with special features, comes with a wide array of accessories.
• No batteries required.
• Some assembly required.
• Recommended for ages 12 years and older.
• Ctn. Pk. Pcs. 6/Ctn. Wgt. Lbs. 29/ Cu. Ft. 5.0
© 1982 Bandai America

62

COMBATTRA™
GoDaiKin™
An Innovative Series of Super Robots
Item #77053
• COMBATTRA™, a super magnetic robot of great force, stands an impressive 11¾" tall.
• COMBATTRA™, the Mighty Battler against all evil powers, is formed when five pieces of space age machinery are united.
• Pull apart COMBATTRA™, and he becomes five military vehicles: Battle Marine, Battle Tank, Battle Kulaft, Battle Jet and Battle Clasher.
• COMBATTRA™ can also be transformed into a space age tank with special features.
• COMBATTRA™, made of durable die cast metal and high impact plastic, comes with various accessories and battle gear.
• No batteries required.
• Some assembly required.
• Recommended for ages 12 years and older.
• Ctn. Pk. Pcs. 6/Ctn. Wgt. Lbs. 25/ Cu. Ft. 5.0
© 1982 Bandai America

63

64

(64) Takara's Diakron Robot Cars of the Future. Vehicles change to fierce Diakron Battle Robots. Diakron Commander ($1^1/_2$ inches tall) and his battle accessories come with each robot car. Commander drives both vehicle and robot. Each of the three robot cars—DK-1, DK-2, DK-3—is made of plastic and die-cast. Retail price, $10 each. *Photo courtesy of Diakron Robot Car of the Future from Takara.*

(65) Armatron, by Tomy Corporation of Carson, California. *Photo by Maggie Steber, courtesy of Robotorium, Inc.*

CONSUMER INFORMATION:
By rotating the joystick controls, Armatron will give the exact motion needed to grip, pick up, rotate, move, and release materials. It comes with canisters, modules, cones, and globes. Requires 2 "D" batteries (not included). Price: $54 at The Robotorium.

65

(66) An adult version of Armatron: Armdroid, a low-cost development tool that can be used in the home, school, factory, or research laboratory as an educational device. Kit $875; assembled $995. *Photo courtesy of Colne Robotics, Inc.*

CONSUMER INFORMATION:

Armdroid is available with two distinct modes of control—computer or manual. The Armdroid can be driven by most microcomputers and can be used as a handling device or alternately as a computer peripheral. All the well-known names will operate the machine, such as Pet, Apple, TRS 80, ZX81, RML 380Z, Acorn, BBC Computer. The bright orange mechanical arm has six degrees of motion and a lift capacity of ten ounces. Ajit Channe, president of Colne Robotics, which produces the Armdroid 1, says it is popular with universities, corporations, and with hobbyists. For further information on the Armdroid, contact Colne Robotics, Inc., 207 N.E. 33rd Street, Fort Lauderdale, Florida 33334. Tel.: (305) 566-3101.

(67) The Rhino XR-1 robot: an educational and research tool for robotics investigations. *Photo and copyrighted promotional material courtesy of Rhino Robots, Inc.*

66

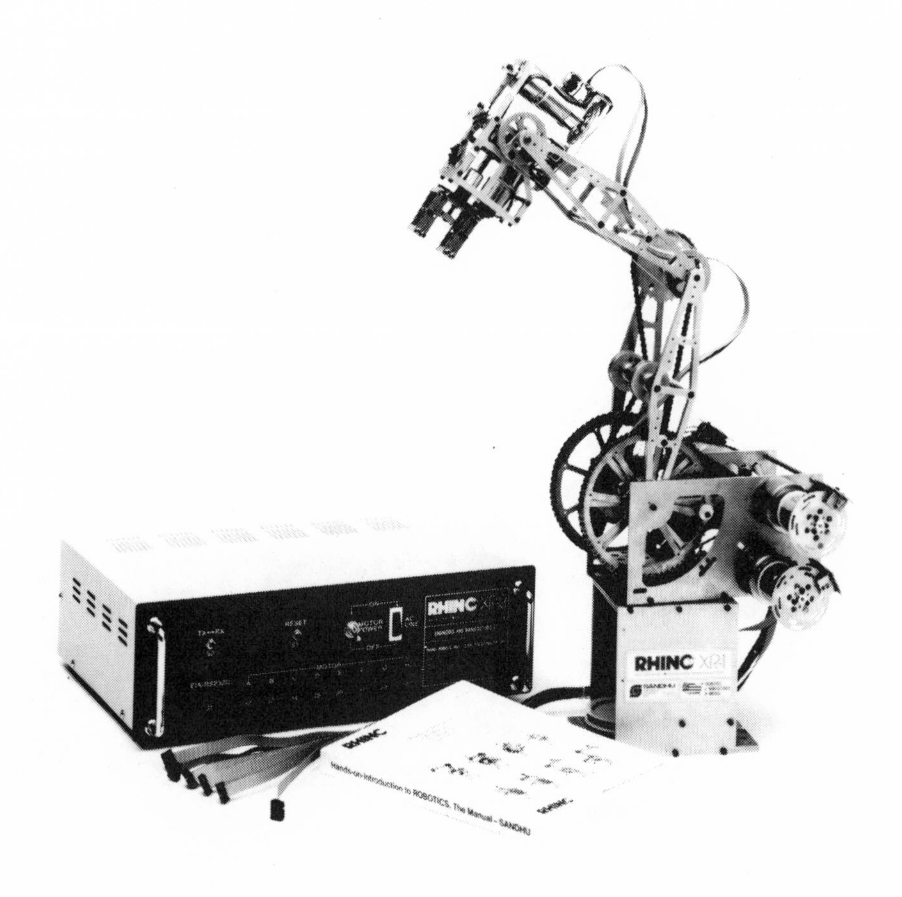

67

CONSUMER INFORMATION:
The Rhino XR-1 Robot is a table-top unit 32 inches high; total weight, approximately 25 pounds. It duplicates the motions and signals used by most industrial robots. Includes standard hand with two-inch fingers, power supplies, eight-axis controller, user's manual, and tool kit. Completely open and observable, the robot has components that can be taken apart and reassembled at the user's option, providing hands-on experience for developing applications adaptable to large robots. The Rhino XR-1 Complete Robotic System is backed by a one-year limited mechanical warranty and a 90-day limited electronic warranty. The system comes complete and ready to operate with any computer having RS-232C interface (three wires). The price of the complete Rhino XR-1 is $2,400, shipped prepaid anywhere in the continental United States. All domestic shipments will be made FOB Champaign, Illinois 61820. Rhino Robots, Inc. normally ships by UPS in the continental US wherever UPS makes deliveries. All other forms of delivery are arranged as requested by the customer. All shipping charges will be collect or prepaid and added to invoice. To order, write or call Rhino Robots, Inc., 308 S. State Street, Champaign, Illinois 61820. Tel.: (217) 352-8485.

(68) Denby is a five-foot-tall promotional robot, leased by the hour, day, week, or month through World of Robots Corporation of Jackson, Michigan. *Photo courtesy World of Robots Corporation.*

CONSUMER INFORMATION:
Denby addresses audiences in space-age speech patterns, and carries on two-way conversations. He has head rotation, elbow and shoulder extension, and hands that grasp. Denby even bends at the waist. The robot is completely wireless and is controlled by professionally trained operators. The World of Robots Corporation advises: "During conventions or trade shows, a representative of your company can help Denby's remote-control operator identify sales prospects and important customers so Denby can talk to them and lead them straight to your exhibit." Denby's rate is $150 an hour. World of Robots offers special rates for long-term leasing. A 40 percent deposit is required of a renter and the balance is paid the day of the show.

Denbys are not offered for sale, but its manufacturer will create custom-made robots. "The least we could quote on custom-made robots," says Joe Collins of World of Robots, "is $7,500. That's for a three-foot robot with an audio system and full mobility. That means he'd have two moving shoulders, one gripping hand, and head rotation. A 50 percent deposit is required at the time of ordering, the balance due on delivery."

World of Robots has established Denbys in most major markets throughout America to cut down on excess charges for transportation, hotel, and other travel expenses. Where there is no branch office, travel expenses are added to the rental fee. For the location of the nearest Denby, and other information, call World of Robots toll free (800) 248-0896. Michigan residents call (517) 788-6840. Written inquiries should be directed to World of Robots Corporation, 2235 East High, Jackson, Michigan 49203.

68

(69-70)
Pictured on opposite page are "The King," Pizza Time Theater's robotic rock 'n roll lion **(69)**—he sings Elvis Presley tunes—and the Pizza Time Players **(70),** who trade jokes and sing in the dining room at Pizza Time Theaters. *Photos courtesy of Pizza Time Theater.*

CONSUMER INFORMATION:
Problem: What to do while patrons are waiting for their pizzas to be cooked? For Nolan Bushnell, who founded Chuck E. Cheese's Pizza Time Theater, the answer was to put on shows performed by robotic entertainers, created in the form of animals. Bushnell launched Pizza Time Theater in May 1977 in San Jose, California. The concept spread. Now a chain of about 250 Pizza Time Theaters operates throughout the U.S. and abroad.

Pizza Time's animation system is the result of a $2.5 million development effort, centering on extensive use of computers. It takes five hours of programming for every one minute of recorded skit material. A unique feature of this system is the automatic volume control. The noise level of the surrounding area is analyzed and the audio level of the show is adjusted automatically. Background music can also be played between skits, and announcements can be made. The character mechanism is composed of a variety of products. Much of the skeleton is aluminum for its light weight and structural integrity. Oil impregnated bronze bearings are used at pivotal points for ease of assembly and maintenance. Pneumatic cylinders of different lengths and diameters are used to create the characters' movements. Some of the parts are unique in their use but not in their design. For example, the eyes are made from drilled-out croquet balls, and the tail pivot is actually a ball joint from an American Motors Pacer.

Shows at Chuck E. Cheese's Pizza Time Theater run every eight minutes throughout the day. For the Pizza Time Theater nearest you contact Pizza Time Theater, Inc., 1213 Innsbruck Drive, Sunnyvale, California 94086. Tel.: (408) 744-7300.

69

70

(71) DC-1, The Drink Caddy Robot, created by Android Amusement Corporation in 1980. *Photo courtesy of Android Amusement Corporation, 1408 East Arrow Highway, Irwindale, California 91706.*

(72) DC-2, from Android Amusement Corporation. *Photo courtesy of Android Amusement Corporation, 1408 East Arrow Highway, Irwindale, California 91706.*

71

CONSUMER INFORMATION:
The creator of DC-2, Gene Beley, says friends of Hugh Hefner gave the *Playboy* publisher the robot as a gift. More routinely, the DC-2 sells to corporations.

DC-2 has a turning head inside a protective dome and its basic model includes the following: HME broadcast microphone for "voice" of the robot; AM-FM radio with auto-reverse cassette, 8-track, or choice of two remote-controlled 8-tracks for sound effects; wheelchair heavy duty motors with pull-out hub caps for free-wheeling base motors; two 12-volt sealed batteries approved for air cargo travel; sequencing L.E.D. lights; foam-padded bumpers; sliding tray; and side-pod cavity toolboxes for miscellaneous items.

DC-2 can utilize a color video camera ($900 extra) in his head, a 9-inch color TV ($500) in his chest and a video cassette recorder ($1,200). Android Amusement Corporation also sells a talk-back "spy" microphone that can be placed inside the DC-2, and a hearing-aid receiver microphone. For business firms that think of the DC-2 as a commercial prop, Beley offers an electronic advertising sign ($495), approximately 12 inches long, mounted on the DC-2's tray.

DC-2 is available through Android Amusement Corporation, 1408 East Arrow Highway, Irwindale, California 91706. Tel.: (213) 303-2434. Basic robot (which excludes color video equipment), $9,500. All prices are F.O.B. Los Angeles, California. All C.O.D. charges and Letters of Credit costs are to be borne by purchasers of robot. A 50 percent deposit is required for Android Amusement to begin

72

construction. Completion of robot takes 45 to 90 days from date Android Amusement's bank accepts deposit funds.

Android's "Purchase Agreement" stipulates: "Purchaser agrees not to enter into manufacturing of any type of radio-controlled or computer robots, or subcontract to any other third party to manufacture robots substantially similar to products manufactured by Android Amusement Corporation; understanding if this does occur, Android Amusement Corporation is entitled to damages of $1,000 per robot for each one sold. . . . In the event purchaser of the robot desires to sell the robot in the future, said purchaser hereby agrees to offer Android Amusement Corporation first right of refusal privileges on purchasing the robot for resale or its own usage."

Android Amusement rents DC-2 for $650 a day in Los Angeles County; $750 (and expenses) outside of Los Angeles County.

(73-74)
Andrea **(73)** and Adam Android **(74),** available for leasing through Android Amusement Corporation at $650 a day in Los Angeles County; $750 a day (and expenses) outside Los Angeles County. *Photos courtesy of Android Amusement Corp., 1408 East Arrow Highway, Irwindale, California 91706.*

CONSUMER INFORMATION:
Andrea Android, 36-24-34 and 5′9″ tall. Her head turns, her arms move. She has complete mobility with no connecting cords. Andrea's audio system includes a wireless mike system for conversations with humans, and an audio tape deck for music in the battery-operated system.

Adam Android, 6′ tall, with a macho California sun-tanned look, is completely mobile. He features a turning head, moving arms, and an audio sound system for voice and music.

73

74

(75-76)
Commander Robot, a robot skater with Ice Follies, is shown here with his creator, David B. Colman, in 1966 **(75)** and, in other shot, as a pairs skater **(76).** Photos courtesy of The Robot Factory.

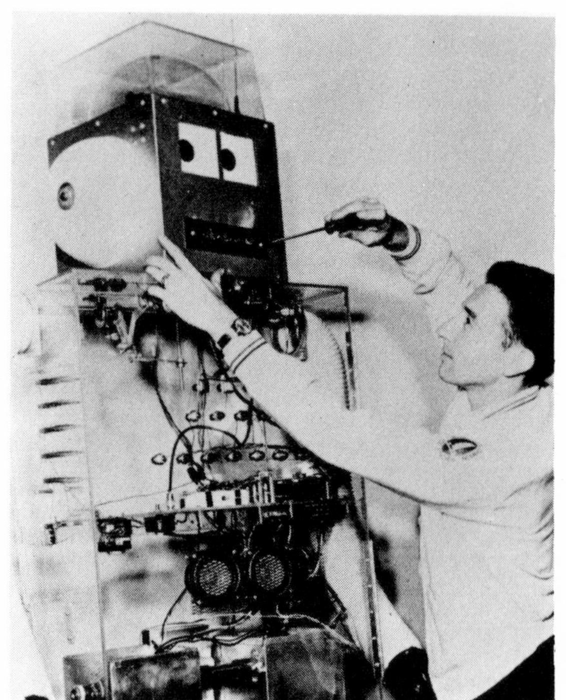

75

76

(77) In September 1982 Six T Robot was stolen. As a result, this poster was disseminated. *Photo by Phil Berger, courtesy of The Robot Factory.*

(78-81)
The Robot Factory sells approximately 100 robots a year—mostly to amusement parks, restaurants, shopping malls, radio-TV stations, and private individuals who start rent-a-robot businesses. Here are several of the standard robots The Robot Factory offers: Ralph Roger Robot, Sr. **(78),** Videobot **(79),** Humanoid Robot **(80),** and Hot Tots—a cross between robots and animated characters **(81).** *Photos by The Robot Factory, Copyright © 1980, 1981.*

CONSUMER INFORMATION:
Ralph Roger Robot, Sr., 7′9″ tall. Walks forward and backward, does spins on his feet. Bends forward at the waist and ankles, and arms move at the shoulders. Head moves from side to side. Talks by

$1500.00

REWARD

For information leading to the arrest and conviction of the person or persons involved in the theft of
Six T. Robot
in Dallas, Texas, on September 13, 1982.

Please Direct All Replies to:

THE ROBOT FACTORY

P. O. BOX 112 CASCADE, COLORADO 80809
(303) 687-6208 (303) 687-6244
ATTN: Margaret Truitt

77

78

wireless microphone. Framework is aluminum and plexiglass (some clear, some mirrorized, and some smoke-colored). Lights flash in chest and eyes. Ralph Roger is a dual-drive robot with extra heavy duty Bodine gearhead 24-volt DC motors. He has two 12-volt gel batteries that provide four to six hours of running time. His power switch is concealed on his body. His controls consist of a self-contained system small enough to hide in a shoulder bag. Price: $14,995. Also available: Ralph Roger Robot, Jr. (5′10″ tall). Price: $9,995.

Videobot is a variant of The Robot Factory's popular Six T Robot. As Six T does, Videobot has flashing lights under his dome and on his chest. The robot sings, dances, plays music, blows up balloons, dispenses cards and brochures, tips its hat, opens and closes its hands. Videobot—priced at $10,495 compared to Six T's $7,995—comes with video monitor and recorder/player. Also available: Videobot with video monitor, recorder/player, and camera ($12,495).

Humanoid Robot is about 5′ 6″ tall, 150 pounds. The robot's head is thin, flexible rubber. Its framework is aluminum with molded plastic, transparent except for legs, feet, and arms. The Humanoid has lights in its head and torso, and has facial and head movements like humans. It walks and talks. Humanoid is a dual-drive robot with low current drain, permanent magnet motors. Two 12-volt gel batteries provide four to six hours of running time. The power switch is concealed on the body. As an option, The Robot Factory can custom mold the head to look like a particular human. Controls consist of a self-contained system small enough to hide in a shoulder bag. Price: $9,995.

Hot Tots ride a variety of children's toy vehicles. When a Hot Tot speaks, its mouth movement is synchronized with its words. Hot Tot bodies are made mostly of fake fur. Power unit varies, depending on what vehicle the Hot Tot is riding. As options, additional bodies and/or heads can be ordered, and costumes as well. Price: $4,500.

All robots from The Robot Factory come with a six-month warranty covering parts and workmanship. Operator training either at The Robot Factory or at purchaser's home is available for a fee. Repair service is provided at the customer's expense. A 50 percent deposit at the time the sales agreement is signed is required, with the remaining 50 percent on delivery. The purchaser pays all shipping and C.O.D. fees.

79

The Robot Factory's profits are mostly through sales, but it also rents its robots. Says Mary J. Bolner, a vice-president at The Robot Factory: "The rental rate varies, depending on the model and how long you want it for. The rental fee includes the robots and its operator. For, let's say, a day or two of renting Six T, the cost would be $600 a day. For three to five days, it's $550 a day. Twenty-five percent is due on signing the contract."

In addition, The Robot Factory does custom-made robots. A unique model—called the Jock-bot—was built for an unidentified race horse owner who has used it to exercise his horses. "It works," says Bolner, "just like an exercise boy. It manipulates the reins, and applies pressure on the sides of the horse. From a distance, the human controls the horse through the robot. Jock-bot is a plastic-covered box, mounted on a saddle. It gives off pressure and has a human-like voice."

For information and sales agreements contact The Robot Factory, PO Box 112, Cascade, Colorado 80809. Tel.: (303) 687-6208 or (303) 687-6244.

80

81

(82) Sico, from International Robotics, Inc. *Photo courtesy of International Robotics, Inc., New York, N.Y.*

CONSUMER INFORMATION:
International Robotics Inc. leases its Sicos to major companies. Price is on a sliding scale, tied to the number of days per year a client employs the robot. "It varies," says International Robotics' president Robert Doornick, "from between $2,000 a day to $3,500 a day." The $2,000 rate is based on an obligation of at least thirty days annual use. Rental price is for an eight-hour day. After that, overtime rates apply. To confirm a rental date for Sico, a one-third deposit must accompany the contract. Two weeks before the rental date, another third of the rental fee is due, along with air fare. The balance is payable after the show. Expenses for travel, $55 per performer per day, and hotel accommodations are additional. Special rates apply for TV commercial and print advertising.

(83) Guide robot for park. **(84)** Beverage robot: dispenses beverage. **(85)** Casino robot. (Artist's renderings by George Hayward.) *Photos courtesy of International Robotics, Inc., New York, N.Y.*

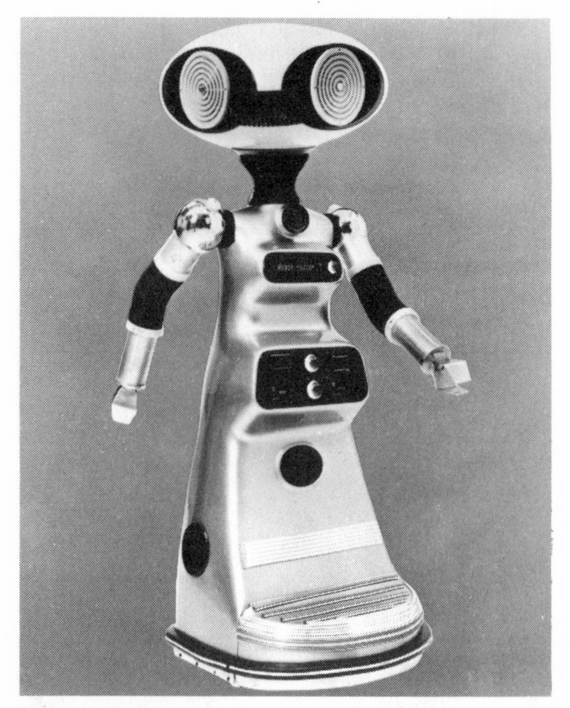

82

83

84

85

(86) Quadracons from ShowAmerica, Inc. *Photo courtesy of ShowAmerica, Inc., Elmhurst, Illinois.*

CONSUMER INFORMATION:
Quadracon is ShowAmerica's standard promotional robot. On occasion it has been adapted to play goalie against puck-shooting fans between periods of hockey games, and once defeated a backgammon champion at his own game. Quadracon stands four feet tall, and is constructed of Fiberglas. Space is provided for a company's logo or sales message. Custom colors are available and the robot can be dressed in special theme attire. Quadracon moves wirelessly throughout exhibit or stage areas, playing recorded messages or music and emitting electronic sounds. It has random flashing lights on the chest and sound-responsive lights in the face panel. Quadracon shakes hands, holds up brochures, premiums, or product samples. The robot is remotely controlled through a two-way FM radio system by a nearby concealed operator. Quadracon can speak different languages or have a female voice for special applications. The robot has a drive system that can handle a variety of level surfaces, indoors or out, and its battery-powered motors run all day with an overnight recharging. Quadracon is compact enough to travel by air at inexpensive baggage rates. ShowAmerica will lease Quadracon for $850 per day, or $3,800 per week, including operator. Layover days at a show site are billed at $250 per day. ShowAmerica also bills for routine expenses (meals, lodging, and transportation). Special rates are available for multiple bookings or time periods extending beyond one week. Rates for foreign bookings are quoted on request.

86

Inquiries regarding Quadracon should be directed to ShowAmerica, 841 North Addison Avenue, Elmhurst, Illinois 60126. Tel.: (312) 834-7500. ("Quadracon" is a trademark of ShowAmerica Inc.)

(87) ShowAmerica Inc. will custom design robots to any shape or character. A shell reflecting the company's product is placed over the motor-box-on-wheels unit that serves ShowAmerica's robots. This unit, called the Robacon (tm), contains the propulsion device, radio controls, and animation equipment. *Photo courtesy of ShowAmerica, Inc., Elmhurst, Illinois.*

CONSUMER INFORMATION:
A custom-designed robot carries a one-time construction charge. Heinz, for instance paid $14,000 for the shell of its ketchup bottle, according to ShowAmerica. The specially created shell remains the property of the client for future shows, but ShowAmerica keeps the Robacon.

"In certain cases," says Wil Anderson, president of ShowAmerica, "we've committed to long-term leases where we've trained the client's personnel in the mechanics and theatrics of the robot and let the robot stay out on the road. But in those cases we're protected by a thirty-page contract that has prohibitions about opening up and examining the Robacon. In those situations, the robot goes out with lead seals. Needless to say, we're being protective about our equipment."

87

(88-91)
Robots custom-made by ShowAmerica:
(88) HJ I and HJ II for Heinz Tomato Ketchup. **(89)** The cartoon character "Dust" for Hoover Ltd, London. **(90)** "Disco" for Time Inc.'s science news magazine *Discover.* **(91)** "Expo Ernie," a spaceman robot—the symbol of the 1986 World's Fair in Vancouver. *Photos courtesy of ShowAmerica, Inc., Elmhurst, Illinois.*

(92) A line drawing of ShowAmerica's newest robot creation, "Peeper." *Photo courtesy of ShowAmerica, Inc., Elmhurst, Illinois.*

CONSUMER INFORMATION:
Peeper is a humanoid robot that stands 56 inches tall but extends to 6 feet with its periscoping head and neck. In addition to extending its neck to peer at spectators, the robot's head moves left and right, and tilts up and down. Featuring wide eyes and voice-activated multicolored lights in its face panel, Peeper can move freely and wirelessly about an exhibit area at trade shows, sales meetings, fairs, and promotional events. The robot is constructed from Fiberglas and has complete arm movement. Rental: $900 to $950 a day.

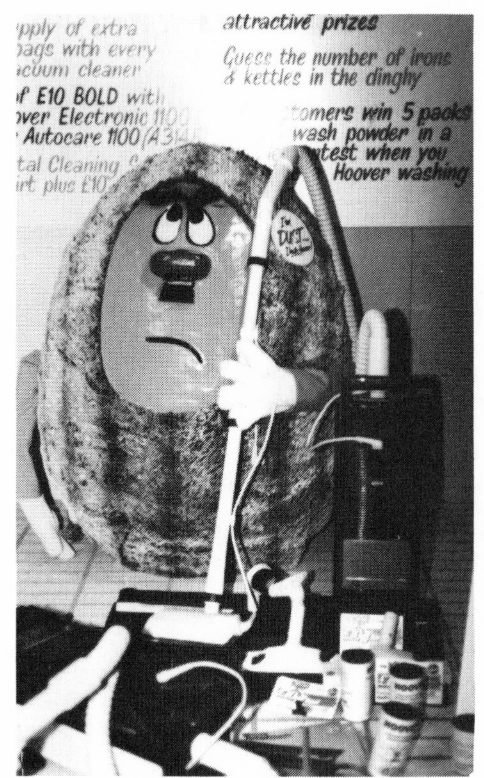

89

88

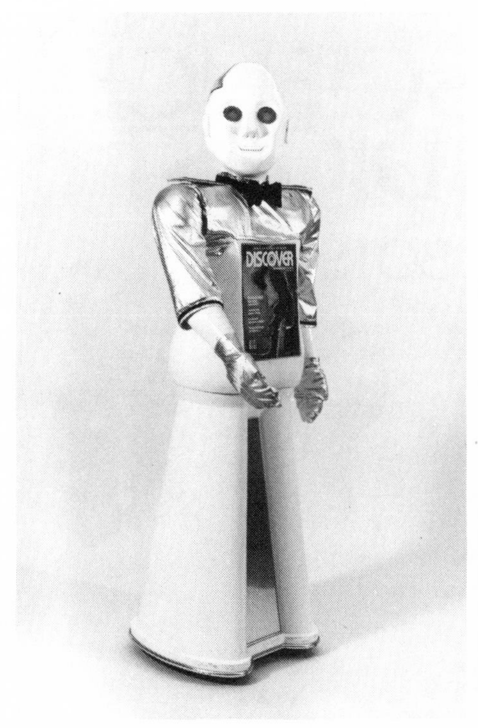
DISCOVER

90

EXPO
86

91

1A
1B
"PEEPER" the Extra Tall Robot to be introduced by
ShowAmerica Inc. in 1983/1984 - Copyright pending
on likeness.

92

(93) The cover of *Science Digest*'s special issue on robots was typical of the extensive media coverage that the introduction of home robots in 1983 inspired. Illustration by Joe Lapinski in *Science Digest,* April 1983. *Photo courtesy of Science Digest.*

CONSUMER INFORMATION:
For robotic developments and other science news, interested parties may subscribe to *Science Digest* by calling this toll free number: (800) 247-5470. Price: $13.97 for 12 issues.

While *Science Digest* covers a spectrum of scientific subjects, there are newsletters and magazines that are more singlemindedly robotic. For subscription information and order forms, write the following:

Robotics Today, One SME Drive, PO Box 930, Dearborn, Michigan 48121. *Robotics Today* is published six times a year by the Society of Manufacturing Engineers in cooperation with Robotics International of SME and the Robot Institute of America. Subscription rates, U.S. and possessions: SME and RIA members $7.50; nonmembers $36.00. Library subscriptions: $12.00. Add $2.00 to rates for Canada and Mexico; add $8.00 to rates for all other countries. For foreign air delivery, contact SME for applicable rates.

Smart Machines, PO Box 459, Sharon, Massachusetts 02067. A one-year subscription (12 issues) is $48. Contact the publisher for foreign and airmail rates.

Robotics Age Magazine, PO Box 358, Peterborough, New Hampshire 03458. Subscriptions are $15 for one year—6 issues. In Canada and Mexico, $17 for one year. In other countries, subscriptions are $19 for one year surface delivery. Air delivery to selected areas is at additional charges, rates on request.

Technology Trends Newsletter, DM Data, Inc., 6900 E. Camelback Road, Suite 700, Scottsdale, Arizona 85251. One year subscription (12 issues) is $295 in the U.S. and $345 (prepaid) for foreign subscriptions.

Industrial Robots International, Technical Insights, Inc., 158 Linwood Plaza, PO

93

Box 1304, Fort Lee, N.J. 07024. Published twice monthly. One year subscription is $192, $48 additional for foreign airmail service.

Robomatix Reporter, 48 West 38th Street, New York, New York 10018. Twelve monthly issues for $250. Overseas clients add $15.

For those involved in industry, further robotics information is available through two professional organizations.

The Robot Institute of America (RIA), founded in 1974, is a trade association serving the industrial robotics field. From an RIA release: "As a trade association, company membership in RIA is open to robot manufacturers, distributors, users, potential users, researchers and accessory suppliers. Over 190 major companies and organizations belong to RIA and receive corporate benefits. . . . RIA provides position papers, a newsletter (*The RIA Monitor*), trade shows and expositions, market studies, workshops, special projects, committee activities, national and international tours, publications, etc." Annual dues range from $200 to $1,000, depending on the class of membership. For further information, contact Robot Institute of America, One SME Drive, P.O. Box 930, Dearborn, Michigan 48128. Tel.: (313) 271-0778.

The Society of Manufacturing Engineers (SME) is a worldwide professional organization for the continuing education of manufacturing managers, engineers and technologists. From an SME release: "Founded in Detroit in 1932 as the American Society of Tool Engineers, SME has grown into an international society of more than 70,000 members in 65 countries. Annual activities include more than 30 combined conferences and expositions, over 250 manufacturing clinics, seminars and symposia, and special interest meetings devoted to such technologies as Industrial Robots and Automation. SME has adapted to the changing needs of the manufacturing community by establishing four specialized associations geared to high technology interests: Robotics International (RI/SME), Computer and Automated Systems Association (CASA/SME), Association for Finishing Processes (AFP/SME), and the North American Manufacturing Research Institution (NAMRI/SME)." Annual dues: $50 first year and $35 thereafter. Fulltime students: $8 per year. For further information, contact Society of Manufacturing Engineers, One SME Drive, P.O. Box 930, Dearborn, Michigan 48121. Tel.: (313) 271-1500.

(94) Androbot's dawning of the robot age—from its promotional literature. *Photo copyright © 1983 by Androbot, Inc.*

94

(95) Heath Co. presented its Hero-1 as an educational tool. *Photo courtesy of Heath Co.*

(96) 1983 was the inaugural year for home robots to appear on the market. An example: RB5X, produced by the RB Robot Corporation, acquired an optional arm, suitable for fetching items like the *Wall Street Journal.* Cost: an additional $595. *Photo courtesy of RB Robot Corporation, Golden, Colorado.*

(97) Nolan Bushnell, of Androbot, Inc., is shown with the company's line of robots. *Photo copyright © 1983 by Androbot, Inc.*

CONSUMER INFORMATION:
Androbot, Inc. has variously priced robots. Topo sells for $795, and is an extension of the home computer. The more sophisticated B.O.B. goes for $2,995. F.R.E.D., under $300, is an educational tool for the children's market.

Androbots are sold through computer dealers and department stores. For the nearest location, contact Androbot, Inc., 1287 Lawrence Station Road, Sunnyvale, California 94086. Tel.: (408) BOBTOPO.

(98) Hero-1 is a product of Heath Co., Benton Harbor, Michigan 49022. *Photo courtesy of Heath Co.*

CONSUMER INFORMATION:
Credit card orders (Visa and MasterCard) for Hero-1 ($1,499.85 in kit form, $2,499.95 fully assembled. Both models include arm and speech synthesizer. A kit form Hero-1 without those options costs $999.95) may be placed by phone through this toll-free number: (800) 253-0570. Alaska, Hawaii, and Michigan residents, call 616-982-3411. The Heath Co. has representatives in all fifty states, and offers a "no obligation" demonstration of the robot. For the name and address of the representative nearest you, call Heath's advertising department at 616-982-3210.

95

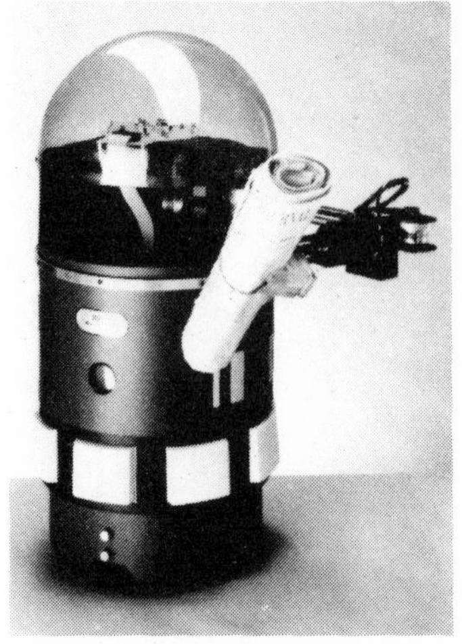

96

97

98

(99) RB5X, with vacuum attachment. *Photo courtesy of RB Robot Corporation, Golden, Colorado.*

CONSUMER INFORMATION:
RB5X, which retails for $1,495, has been on the market since January 1983 and is available through a growing network of computer dealers across the United States, and in Canada, the Netherlands, Japan, and Colombia. For the name of the closest dealer or for more information, contact RB Robot Corporation, 14618 West 6th Avenue, Suite 201, Golden, Colorado 80401. Tel.: (303) 279-5525. RB5X can also be ordered by mail. Payment by check or credit cards (Visa/MasterCard/American Express). Colorado residents add 3 percent sales tax. Allow sixty days for delivery. Write or phone the manufacturer for an order form.

(100) Genus. *Photo courtesy of Robotics International Corporation.*

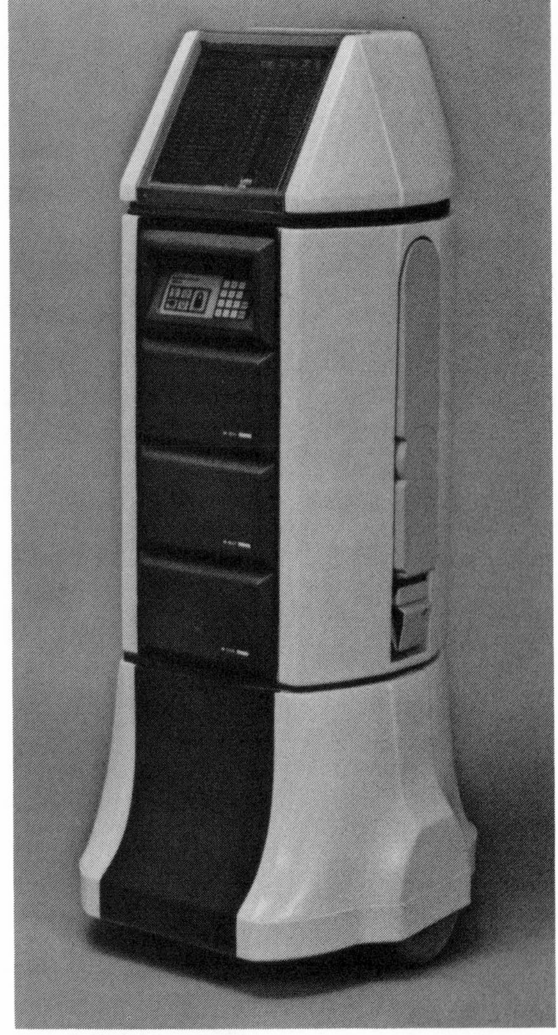

100

99

CONSUMER INFORMATION
Genus, a product of Robotics International Corporation, was expected to be mass-produced by early 1984 and sold through a nationwide network of independent home computer stores with in-shop capabilities for servicing the robot. Robotics International anticipated charging $10,000 for Genus, the price to include a vacuum sweeper and a security package. Joe Collins, director of special projects at Robotics International, says the company plans that sales of Genus will be *exclu-*

sively through their dealers. For latest information on the availability of Genus, and other particulars, contact Robotics International Corporation, 2335 East High Street, Jackson, Michigan 49203. Tel.: (517) 788-6840.

(101) Itsabox. *Photo courtesy of Technical Micro Systems, Inc.*

CONSUMER INFORMATION:
Inquiries about Itsabox should be made to Technical Micro Systems, Inc., 366 Cloverdale, Ann Arbor, Michigan 48105. Tel.: (313) 994-0784. In the summer of 1983, TMSI's Lee A. Hart indicated that the product was in process, and nearing production. "We've conducted a telephone survey with promising results," Hart said, "and used this information to prepare design specs, cost estimates (price: kit version under $500; assembled $600), and a preliminary description. We'll be mailing this out with a survey to our mailing list to determine how best to proceed. At this point, customer feedback is *very* important—there's no future in designing a product that doesn't do what people want. Like the early home computer market before it, the hobby robot market seems bent on creating great expectations and overselling its capabilities. I feel this will disappoint many customers. The educational market looks more rational, but even there the educators often know as little about robots as their intended students."

(102) Line drawing of Companion, the ultimate home robot, conceived by International Robotics, Inc. and available through Sperry & Hutchinson's exclusive catalog only. Price (depending on options included): the equivalent in S&H media of $150,000 to $250,000. *Photo courtesy of International Robotics, Inc., New York, N.Y.*

CONSUMER INFORMATION:
For further information on Companion, or a copy of the company's catalog, contact Sperry & Hutchinson Co., Inc., 330 Madison Avenue, New York, N.Y. 10017. Tel.: (212) 983-2000.

101

102

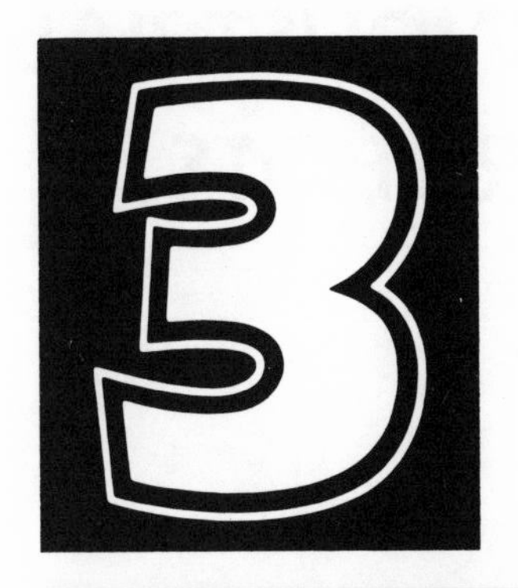

INDUSTRIAL ROBOTS

Plus Miscellaneous Others

INDUSTRIAL ROBOTS

The patents for industrial robots were held long before robots themselves became standard in factories across the world.

It wasn't that industrial robots couldn't be built back in the 1950s when the so-called "father" of these robots, George C. Devol, already had more than forty patents for the basic technology on which the industry is founded. The problem lay with Devol himself. As robot manufacturing pioneer Joseph Engelberger tells it: "George Devol was unable to restrain himself from spilling the whole dream out, which scared most businessmen off."

About the only man Devol didn't scare off was Engelberger. Engelberger thought that the "smart machines" that Devol so fervently espoused were more than the science fiction oddities others made them out to be. Engelberger believed that these machines, these robots, could make a difference in our lives, increasing productivity while decreasing the tedium and dangers that industrial workers faced (103-106).

In 1956, Engelberger started Unimation, Inc., a company that would not turn a profit until 1975—nineteen years after its conception. Back then, Engelberger found, the very word *robot* antagonized businessmen, who reacted as if gimmickry were at the heart of the concept.

Engelberger's effort to make robots part of the industrial landscape was a trial-and-error process, and sometimes progress came from unexpected quarters. For instance, spot welding is an established application for robots today, but in Unimation's early days it was a usage that Engelberger had not even considered. The idea for spot welding would come out of a conversation the Unimation boss had with executives at General Motors.

In those early years, Engelberger's was a voice that went largely unheeded here in the United States. But in Japan, when Engelberger arrived there in 1967 to describe the future he saw once robots were in place, people listened. Indeed, after Engelberger's formal presentation, Japanese executives and engineers in attendance questioned him for five straight hours about robots. A year later, Kawasaki Heavy Industries, Ltd., now Japan's leading robot maker, would begin producing Japan's first robots as a licensee of Unimation, Inc.

By the 1970s, the Japanese had embarked on a policy of robot application that has made them the undisputed robotics leader in the world today. Along the way, the Japanese discovered that the money expended on robots would return dividends through reduced production costs and increased productivity—at least 30 percent according to the Japan Industrial Robot Association. Not only that, but the quality of products improved. The robot almost completely eliminated the element of human error and brought higher reliability; product defects decreased from a 5 percent rate to a 0.1 percent rate. And if robots—contrary to the popular impression—sometimes worked more slowly than human beings, they were able to labor twenty-four hours straight, if necessary, and be reprogrammed the following day for a changed production line. (That ability to be reprogrammed, as well as its possessing a multifunctional manipulator, distinguishes a robot from merely automated machinery.)

The Japanese became the first people to discover what an asset an industrial robot could be. In part, necessity moved them to react more swiftly than the rest of us to the industrial robot. The rise of robots in Japan cannot be separated from the effect the oil crisis of 1973–74 had on its economy. When the oil-producing nations formed OPEC and began to squeeze consumer nations through managed prices, industrial production in Japan dropped 14.5 percent and corporate incomes in nominal yen dropped 30.7 percent. Japan's difficulties were exacerbated by the long-standing policy its major corporations had of not laying off workers. As a result, industry there had to contend with more burdensome labor costs.

In time, managements encouraged early retirement and cut back on new hiring to reduce labor costs. But even after the economy was back on the upswing, corporations there persisted in more efficient use of labor, and robots were the linchpin of their strategy.

Japan's industrial problems were, and are, compounded by a thinning population—the birth rate has dropped from 2.1 children per woman to 1.7—and a reluctance on the part of young people coming out of schools to buck the seniority system that rewards older workers with higher wages: twice as much for personnel over 40 as for those in the 20 to 24 age group. According to a report issued by the Japan External Trade Organization (JETRO): "Demographic trends indicate that the age of the work force—and wage bills—will rise steadily during the 1980s. It will be difficult for companies to rejuvenate their work forces. Not only are there fewer young people coming out of schools, but fewer of the graduates are going into manufacturing jobs. Young Japanese today are seeking higher education and prestigious, high-paying white-collar jobs. The percentage of high school graduates entering college rose to 37.4 in 1979 from 32.2 percent in 1973 and 16.1 percent in 1966. . . . So Japanese industry faces a gradual decline in the number of workers during the 1980s. But as robot production continues to grow and robots are adapted to a greater number of manufacturing processes, industrial production is expected to continue rising throughout the decade."

Whether the future turns out that bright remains to be seen. But given the way Japan so far has incorporated robots into its work day, JETRO's roseate picture is not unrealistic. In expanding the role robots have played, Japanese industry has traded on the good will its workers have had for them. Employees are known to say *"Ohayo gozaimasu"* ("Good morning" in English) to robots at the start of a day. And, as Henry Scott Stokes reported in the *New York Times Sunday Magazine*: " . . . the Japanese love affair with robots dates back decades and, in the view of many authorities, is a unique, intensely personal reaction with roots in Buddhist values." Japanese psychologist Seiichiro Akiyama told Stokes: "We give them [the robots] names. We want to stroke them. We respond to them not as machines but as close-to-human beings."

Official Japan has been more shrewd than sentimental in its view of robots. Early on, the government formulated policies regarding robots that propelled the nation headlong toward profit. An important move was to make robots accessible to small enterprises that in ordinary circumstances would not have had the capital to buy them. With the aid of the Japan Development Bank, the government established the Japan Robot Leasing Company, a collaboration of robot manufacturers and insurance companies that enabled the small businessman to lease his robots for $90 a month and trade up when improved ones became available. Without tying up large sums in an outright purchase, the smaller Japanese capitalist enjoyed the same cost reductions and production gains that the Kawasakis and Mitsubishis have attained through robots.

Those gains are most apparent in the automobile industry, which was dominated by the United States before Japan gained control of the market. Firms like Toyota and Honda are able to produce cars that the United States Department of Transportation estimates have a

$1,000 to $1,500 a car cost advantage, secured in large part through robots.

Not only did robots lower labor costs in Japan, but they produced a higher quality automobile—a car for which fewer repairs had to be made in the factory before it was sold, and less warranty work done after it was in a buyer's hands. That, in turn, widened the cost advantage and continued the cycle of success Japan's automotive industry has had.

One may reasonably ask why U.S. industry resisted robots at a time when Japan was seizing the initiative. The answer has to do with the philosophical differences with which business is done in Japan and in the United States. In Japan, major companies, as we have seen, do not lay off their workers; a policy of guaranteed lifetime employment exists. That is the starting point of a system which encourages cooperation between labor and management, and a concern for the long-term profit picture.

In the United States, the perspective is different. As American robotics expert Paul Aron says: " . . . the last thing that American managers desire is to see the price of the company's stock go down, making their options worthless. Stock prices will decline if profits are reduced. Therefore, the sizeable investment in equipment and plant, and the major reorganization and production interruption often required to install robots makes American managers very hesitant to introduce that technology. Furthermore, many American managers are nomadic or migratory, working for a company for a certain number of years and then moving on to another company. Under these conditions, a manager will not introduce robots when they would result in profits only after he has left the company, three, four, or five years later. The manager's outlook in America tends to be short-range, while in Japan it tends to be long-range."

Japan's outlook has resulted in far-ranging robotic applications, including the once futuristic notion of unmanned factories (107-110). At the base of Mt. Fuji, and at other sites in Japan, unmanned factories have become prime tourist attractions—wondrous examples of what a highly concerted technological effort can do.

The factory at the foot of Fuji is Fujitsu Fanuc, Ltd., the world's largest maker of CNC (computerized numerical control) systems for machine tools and one of the hottest technology-oriented firms in Japan. It is also a showcase of unmanned production. But that striking technology does not occur during daylight hours when the Fuji factory employs assembly workers who produce robots and CNC machinery. It is at night that Fujitsu Fanuc becomes special. Then the machinery takes over. CNC machine tools and robots and automatic pallet change operate during the night shift, with a lone human manning the central control room. In a milieu that seems created by the likes of Isaac Asimov, robots actually make other robots.

In Nagoya, the Yamazaki Machinery Works, Ltd. has been an unmanned facility at night and, as described by *The New York Times'* Steve Lohr, a vision of eerie, antiseptic efficiency:

> What immediately catches the eye is the movement of the machines. They do not perform in unison, which is the characteristic pattern of traditional automation. Rather, each machine works independently, making an individual part different from its neighboring machining center. The computer tells a machine tool to drop one task, pick up another, speed up, slow down or whatever—all in sync with the overall, computerized production plan.

Nothing quite so elaborate has yet appeared in the United States. The robots here are fewer and more conventional in their applications. But they stand up in quality (111-112) to Japanese robots. Industrial analysts here insist that Japan's edge in robots is not due to technological innovation but to its skillful application. As Unimation's Engelberger says: "I still haven't seen anything outstanding technologically over there."

In the marketplace today, robots are available that can fit many needs and satisfy the budgets of a major company or a small shop (113-127). At Chicago's Robots 7 Exposition, scores of companies displayed robots ranging in price from several thousand dollars to $100,000 plus.

In the early stages of factory automation, robots were often single-task machines of the nonservo variety (meaning noncomputer-controlled) and were capable only of linear motion. Typically referred to as "pick and place" robots, they were basically mechanical arms capable of materials transfer. They operated through a series of preset electronic or stop switches. These stops were totally responsible for the robot's ability to work through a job sequence.

For those who stand by the definition adopted by the Robot Institute of America, these nonservo machines do not even qualify as robots. By the purist's view, the earliest robots used for factory automation, called "first generation," are fixed sequence devices. Many of these robots are taught by an operator who literally leads it by the hand from one step to the next. This sequence is then recorded by the robot's control unit for use in production. Once the robots have been set up, or programmed, to perform a sequence of operations, they can repeat this sequence tirelessly and reliably. To do a different job, however, they have to be taught to perform a new sequence of operations.

As George C. Dodd, of General Motors Research Laboratories, says: "The basic limitations of first generation robots are: the part being handled must be in precisely the correct position; the part must be stationary; the robot cannot adapt to changing situations because it cannot see or feel the part."

Until recently, these first generation robots (aka "dumb robots") have predominated, performing such functions as painting, spot welding and loading—tasks that are unpleasant and present occupational hazards.

Second generation robots are controlled by minicomputers which increase their flexibility. They calculate how to move depending on which kind of part arrives at the work station and how the part is positioned. They can also track moving parts on assembly lines. However, these robots still lack the capability to sense and adapt to a continuously changing environment. Adaptation requires that the robot be provided with various sensory inputs.

Probably the most obvious sense is vision. If the robot can see where the part is, it can figure out how to pick it up. A sense of touch can also be very useful in many assembly operations such as screwing bolts into holes. Assembly robots herald the new wave of intelligent robotics systems capable of vision, touch, torque, and force.

Traditional industrial robots have been designed from a machine-tool perspective. They are built for durability and the lifting of maximum payloads, with accuracy not as crucial. Light assembly robots represent a significant departure. They are designed from a computer perspective for use in high technology manufacturing (128-129). In this environment, lifting of weight is secondary to accurate placement and delicate handling of precision parts.

Before development of the microprocessor it would have been technically impossible to build a lightweight, flexible robot capable of delivering extreme accuracy. Now VSLI (Very Large Scale Integrated) circuitry and user-friendly software make such a system possible.

In addition, the trend toward decreasing technology prices is enabling robot manufacturers to begin delivering products with acceptable payback periods. The Robot Institute of America estimates that the cost of robots is recovered within the first two years of use.

Vision has expanded the potential of robots (130-133). As Philippe Villers, president and director of Automatix, Inc., says: "Vision systems permit inspection, sorting, and robot guidance. Other systems perform arc welding (the melting of metal filler to join two parts along a crack or seam) and assembly and testing of products, which require sophisticated control of the robots' motions during the entire process. The arc welding function is very important because a worker must wear a protective mask to perform this job, reducing his productivity. With a robot and human being working together, productivity rises from 10 to 20 percent to 70 to 90 percent."

A highly publicized development in 1983 was GE's sensor system (134) that brings the powers of sight and intelligence to welding robots, enabling a welding robot for the first time ever (according to the manufacturer) to steer itself along irregularly shaped joints, continually observing the joint and weld puddle and making adjustments as it traverses along.

With or without vision, robots are available for prospective users in free-standing, single cell applications units (including hardware and software), or in a series of integrated units (systems) that can be coordinated or interfaced with other advanced manufacturing equipment.

The Japanese are heavily into "integrated application packages," to use the insider's phrase for automated assembly systems that include robots. Nobody doubts that the industrial future lies in that direction (135). And in Japan, the future is now. By contrast, American industrialists have lagged behind, not coming to the robotics race with real fervor until the end of the 1970s. Their late start is reflected in the domestic sales figures for robot manufacturers:

1978—$35 million
1979—$60 million
1980—$90 million
1981—$155 million
1982—$185 million

Contrast that with Japan's sales figures during part of that period. As reported by JETRO: "Robot production in Japan began to surge in 1976, as Japan recovered from the recession of 1974, and has been growing ever since. But the industry is still small; total production in 1980 was only 19,900 units, worth 78 billion yen ($340 million)."

For industrial robotics in the United States 1979 is regarded as a pivotal year. That was when Chrysler Corporation placed an order for 100 Unimate robots that was, according to Unimation, Inc., the largest single sale of robots in U.S. automotive history.

From that point, the major corporations moved aggressively on the robotics market (136-137). Through acquisition (Westinghouse Electric Corporation acquired Unimation, Inc. in December 1982 for a reported $107 million), through joint ventures (General Motors Corporation and Fujitsu Fanuc, Ltd. combined to form GMFanuc Robotics Corporation), and through avowed intention (IBM, General Electric, Bendix), high-powered American corporations became part of the robotics industry, changing that business forever.

Despite anticipated inroads by electronic and light manufacturing industries over the next five years, automobile corporations like Chrysler and General Motors will still be the primary users of robots in the forseeable future (138-139). In 1975, Chrysler had only sixteen robots. By 1981, the company increased that number to 240. For the most part, the robots have been used to automate the body shops in Chrysler's four front wheel drive U.S. car assembly plants. At those four plants, 95 percent of all welds are now done automatically. The furious welding activity at one station of the assembly process has prompted Chrysler employees to refer to it as the "turkey farm" because the robotic weld guns seem to peck incessantly.

Chrysler's master plan calls for expanded use of robots, with as many as 987 in operation by 1988. By that time, Chrysler's robots will be handling these new tasks: fusion welding with seam tracking; mounting wheels and tires; installing engines and axles; loading and unloading stamping presses and die-casting machines; installations of windshields and backlights; application of adhesive sealer.

Like Chrysler, GM also has a "last word" auto assembly plant, which was to begin turning out Cadillacs and Oldsmobiles in 1983. Located in Orion Township, Michigan and built at a cost of $500 million, the seventy-two-acre plant will employ robots in 85 percent of the welding done on new autos, and all the painting. The facility in Orion Township will be one of three new plants GM is building, each costing $500 million to construct. By 1990, GM is expecting to utilize 15,000 robots, up from the estimated 1,500 robots it had in 1982.

MISCELLANEOUS ROBOTS

While work robots by now are routinely associated with industry, their uses go well beyond factory walls. Increasingly, technology has found ways to expand the potential of robots in making them the complete helpmate that man has envisioned from Aristotle's time.

Robots now function in a variety of roles, some of them in situations sensational enough to bring headlines. A bomb carrier robot (140) produced by Pedsco Canada, Ltd. came to light when a New York City patrolman and two bomb squad officers were critically injured in two separate blasts in 1982. Those incidents prompted New York City Mayor Edward Koch to say: "I don't understand why, in this technological age, when you have robots, that you can't use a robot's arm to pick up the bomb." Soon after, the New York Police Department had purchased three remote-controlled bomb carriers.

At a price reportedly between $20,000 and $50,000—depending on accessories—the Pedsco device is 18 inches high and weighs 230 pounds. According to published reports, the robot possesses two remote controlled arms, and one "hand" capable of lifting loads of up to 70 pounds. Through a video camera, an officer can work from a remote position. The RMT-Mark 3, the official nomenclature of the robot, is also fitted with water cannons to soak suspected bombs. X-ray vision is an available option.

In addition to bomb disposal work, the Pedsco robot can be adapted for surveillance, fire fighting, radioactive environments, or hostage situations. It can also be equipped with a shotgun. As Phoenix Detective Bob Horath told the *Robotix News*: "It has a shotgun on it that can be used for whatever reasons you'd want to use a shotgun—from destroying an explosive device to a hostage situation in which you want to secure an area. You can let the machine go first and secure the area to make sure that there's no one waiting in ambush."

Another robot called upon in hazardous circumstances is the Y-12 Plant mobile manipulator, known as Herman. Herman is used in toxic or radioactive environments (141). The manipulator is designed to operate at distances up to 700 feet from its control console, to which it is attached by a cable. It has a mechanical hand capable of lifting 160 pounds and dragging up to 500 pounds. Two television cameras are mounted behind the arm to transmit pictures to monitors on the control console. The manipulator was designed by personnel of Union Carbide to specifications by a commercial vendor in 1966. The manipulator normally is housed at the Oak Ridge Y-12 Plant (Oak Ridge, Tennessee), operated by Union Carbide's Nuclear Division for the U.S. Department of Energy.

The manipulator is 62 inches high and weighs 1,800 pounds. It has been used outside the Oak Ridge area on several occasions to assist in the recovery of radioactive materials. In August 1975, Herman was used to recover a radioactive pellet from an irradiation facility at the University of Rochester. On other occasions, it was put to work in irradiation facilities at the University of the South in Sewanee, Tennessee and in Bethesda, Maryland. In March 1979, the Y-12 plant manipulator was taken to the Three Mile Island nuclear power plant in Pennsylvania for standby duty, but it was not used.

As adept as Y-12 is for emergencies, it is a relative toy compared to robots that the National Aeronautics and Space Administration (NASA) sent to Mars. In July 1976, the Viking I Lander came to rest on Mars (142) and transmitted information and photographs until November 1982 when, inexplicably, it went silent.

The robot craft was one of two Viking Landers that were launched at the time. Each of the Vikings sent to Mars was a double spaceship.

One part was an Orbiter that circled Mars, photographing the surface and analyzing its atmosphere from far above the planet. The other half, the robotic Lander, was programmed to explore the surface of Mars, and radio back its findings about the soil, atmosphere, and chemical make-up of the red planet (143).

The Viking Lander was more technologically complex than any automatic spacecraft launched before. As NASA's own pamphlet, *Mars: The Viking Discoveries,* puts it: "Each Lander looks like a cluttered six-sided workbench with three legs, but it contains the equivalent of two power stations, two computer centers, a TV studio, a weather station, an earthquake detector, two chemical laboratories (one for organic and one for inorganic analyses), three separate incubators for any Martian life, a scoop and backhoe for digging trenches and collecting soil samples, and miniature railroad cars for delivering the samples to the laboratories and incubators. Equipment that would normally fill several buildings has been designed in miniature to fit in a spacecraft less than three meters [ten feet] across."

Future space explorations will rely on robotic devices not only to explore planets, as Viking Lander I did, but to service satellites already in place. Grumman Aerospace, for instance, has on its drawing board (144) a flying drone that could exit from a shuttle craft, and secure a satellite to tow it back for repairs.

The robot aide is being called the Proximity Operations Vehicle (POV) and, as designed by Grumman, is a thirty-cubic-foot box with four spindly arms. One of the problems a shuttle would have in approaching a satellite is the grimy exhaust its thrusters emit when they fire. The sooty gas could contaminate dirt-sensitive satellites. A key feature of the Proximity Operations Vehicle is its use of a noncontaminating cold gas propulsion system to perform its maneuvers. As currently designed, the POV could either be manned or remotely controlled by an astronaut inside the shuttle. Either way, it would work effectively up to one half mile from Orbiter.

Back here on Earth, robots are being tried in every conceivable way, from the Japanese model that makes rice patties for sushi (1,200 an hour, or three times faster than an experienced chef) to the robot mail cart that moves unmanned through office corridors (145), to the chemical laboratory robot called the Zymate (146) that can perform delicate scientific tasks. The Australian wool industry is even using a robot to shear its sheep (147).

Exxon has gone underwater with unmanned equipment to produce oil. "With the Subsea Production System (SPS)," says A. C. Garner, Jr., manager of Exxon USA's Southeastern Production Division, "we produced oil for the first time through a sea floor production system that was totally unmanned in the conventional sense. The SPS was remotely installed, remotely operated, and remotely maintained, without the use of divers."

Research on the SPS began in 1968 when Exxon realized that the search for oil and gas would soon extend beyond the capability of conventional offshore production technology. Construction of the SPS got underway in June 1973 and the template was launched in October 1974.

Using space age deepwater technology, wells were drilled and five pipelines and two power and control cables were laid and connected to the template.

When needed, the so-called maintenance manipulator (148) was taken from its onshore storage site and used by remote control to perform routine maintenance operations. Launched from a workboat, the robotic manipulator descended to the template, where it moved along a track around the unit to the desired location. A system of underwater television cameras built into the manipulator allowed the operator on the surface to watch and control its movements as it replaced the failed piece of equipment on the SPS.

"The SPS project developed and demonstrated a diverless system capable of producing oil and gas in water depths beyond 2,000 feet," says Garner. "In the United States, the search for new reserves is rapidly approaching that depth. Overseas, Exxon has drilled exploratory wells in 4,000 feet of water."

Development of offshore reserves by conventional, fixed platforms is limited, primarily by costs, to water depths of 800 to 1,000 feet. More exotic types of offshore platforms, such as the guyed or compliant tower, are being designed for use in water depths from 800 to 2,000 feet.

Exxon invested more than $81 million in the SPS project. "By the time commercial reserves of oil and gas are discovered beneath the deep waters of the world's oceans, we want to be ready," Garner says.

Crucial work is also being done with robots that can make life easier for the handicapped. But because of the limited market for such devices, sometimes patients' needs can become entangled with the manufacturer's profit motives. Take, for instance, the voice-controlled powered wheelchair that was tried at New York University Medical Center's Institute of Rehabilitation Medicine, a facility for the severely handicapped. According to Ruth Dickey, coordinator of NYU's Electronic Technical Aids Evaluation Project, the manufacturer of the wheelchair was unwilling to produce the conveyance unless it was ordered in lots of fifty, a quantity deemed too voluminous by NYU.

But more than economics weighed against the prototype of the voice-controlled powered wheelchair, which could respond to thirty-six spoken commands. The voice-controlled powered wheelchair had limits that the wheelchair currently used by quadriplegics at NYU (the sip and puff wheelchair) does not. The voice-controlled wheelchair was programmed for only one voice. And while it would respond to that voice, it resisted pressure applied by others trying to help the patient. Moreover, it asked too much of the disabled patients—"too many inputs," Dickey says.

The sip and puff chair—so called because patients move it by breathing into a tube or sipping from it—is less demanding on a quadriplegic and is responsive to parties other than the patient. In addition, it costs far less at about $7,000 than the voice-controlled chair. "With a hard puff," says Dickey, "the chair will go forward. With a hard sip it goes into reverse. Either way it latches so you don't have to constantly sip and puff to keep it moving. If you want to make left and right turns, you do it with *light* sips and puffs."

Another device for the handicapped is the medical manipulator (149-150) that is being developed by Carl P. Mason for the Veterans Administration Prosthetic Center in New York. The medical manipulator is a robotic arm that a quadriplegic can use from a wheelchair. Through pressure applied by the chin or through a computer-linked voice recognition system, the patient can use the manipulator to pick up objects, to make phone calls, to close doors, or even to maneuver the pieces on a chess board.

A device called the AbilityPhone terminal does even more. AbilityPhone (151) not only serves as an accessible telephone for the handicapped, but also can be user programmed to turn out the lights, unlock doors, and control electrical appliances. One woman, according to the manufacturer (Basic Telecommunications Corporation) has even programmed the AbilityPhone to automatically water her plants.

The system originally was designed in response to government pressure on phone companies to provide the handicapped with full and equal access to telephone facilities. But as Alan Brown, president of Basic Telecommunications, says: "People with disabilities such as cerebral palsy, muscular dystrophy, arthritis, or paralysis by spinal cord injury all have different needs. Thus, Basic Telecommunications proceeded with plans for a flexible system that could be user programmed to perform a variety of communications functions and to match the physical capabilities of the individual."

The AbilityPhone replaces a standard telephone and consists of a single-line visual display attached to an easy-to-operate keyboard. An optional voice synthesizer can speak program commands and messages for the visually impaired, and can also function as a nonvocal communications device for the vocally disabled.

The keyboard can either be activated directly or by remote switches sensitive to the slightest pressure from any part of the body. Breath activation switches are also available. As a personal communications system, the AbilityPhone features an emergency call mode that monitors the status of the user by asking, "Are you OK?" at predetermined times. If the user fails to respond, the system will automatically dial preprogrammed numbers and summon help with the voice synthesizer. This emergency call can also be triggered remotely or directly by the individual. In addition, the system can give reminder messages such as, "It is time to take your pill." and can be used as an alarm clock.

CATALOG AND CONSUMER INFORMATION

(103-106)
The Unimate 2000 pictured in die-casting **(103)**; the Kuka IR 601/60 seen in glass-handing **(104)**; Spine's paint robot **(105)**; and the Kuka robot **(106)** that automatically mounts and torgues multiple lugs of an automotive wheel show the diversity of difficult and sometimes dangerous tasks that industrial robots can perform. *Photo 103 courtesy of Unimation, Inc., a Westinghouse Electric Co; 104 and 106 courtesy of Expert Automation, Division of Kuka, Inc.; 105 courtesy of Spine Robotics.*

103

104

105

106

(107-110)
Unmanned factories like the Fanuc robotized motor factory epitomize Japan's technological superiority. The diagram **(107)** of the two-story robotized factory depicts a system that includes unmanned parts carriers, robot machines, assembly loading, and an automated warehouse. The facility produces servo motors for the GMFanuc robot. Other photos reveal the robotized machining system **(108)** loading and unloading a workpiece and, in close-up, a robot removing a part from a pallet for machine loading **(109).** The process that incorporates four Fanuc robots **(110)** ends with an assembled motor. *Photos courtesy of Fanuc Ltd.*

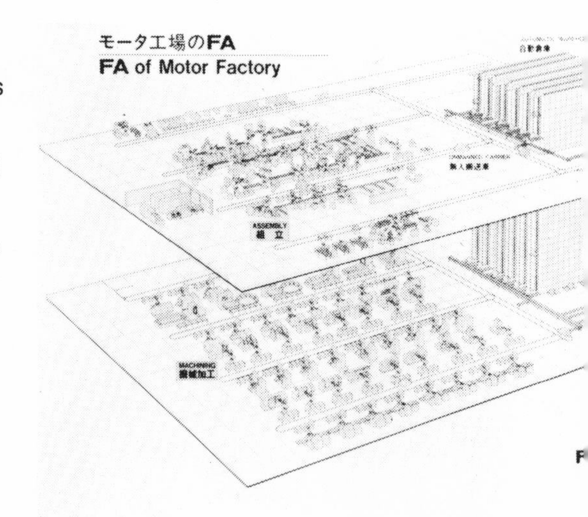

107

108

109

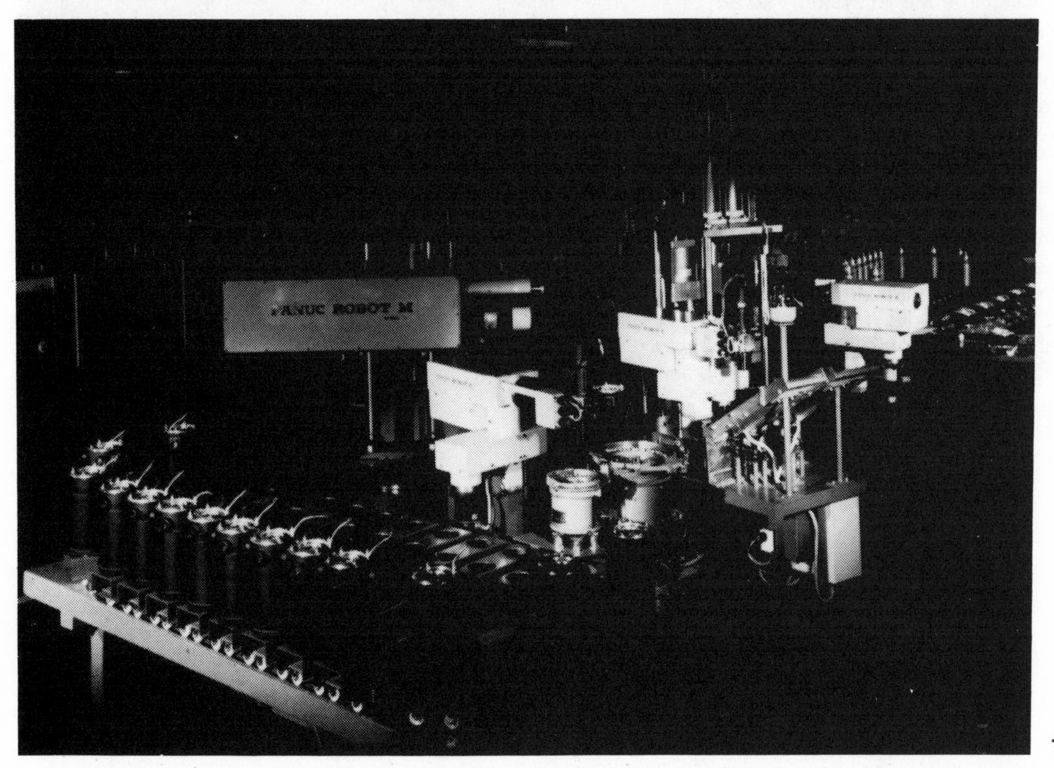

110

(111-112)
The Merlin robot, from American Robot Corporation.

(112) The Merlin is built for precision operations. To show its dexterity, in this instance the Merlin was programmed to thread the eye of a needle 0.009-inches with a needle 0.0007 inches wide. *Photos courtesy of American Robot Corporation.*

CONSUMER INFORMATION:
In March 1983 the American Robot Corporation of Pittsburgh began shipping its Merlin Robot to Marubeni Corporation of Japan. "We believe," says Romesh Wadhwani, president and chief executive officer of American, "this is the first time that advanced flexible manufacturing robots have been exported to Japan." The total value of the Japanese purchase was more than $250,000 for five robots. Marubeni will distribute Merlin robots in the Japanese market for use in high speed materials handling, precision electronics assembly, testing, and arc welding. The Merlin system can be expanded step by step from a single robot to a flexible manufacturing cell to a fully automated factory, a capability that has strong interest in Japan, according to Wadhwani. American stresses the Merlin's ability to repeat tasks to within tolerances of ±.0001-inch. That quality combines with payload capabilities up to fifty pounds and arm speeds in excess of five feet per second. For further information, contact American Robot Corporation, 121 Industry Drive, Pittsburgh, Pennsylvania 15275. Tel.: (412) 787-3000. *Photo courtesy of American Robot Corporation.*

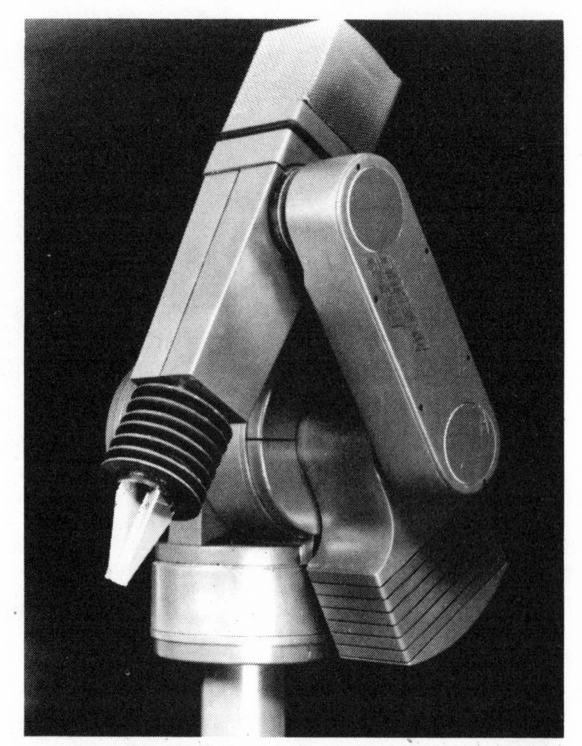

111

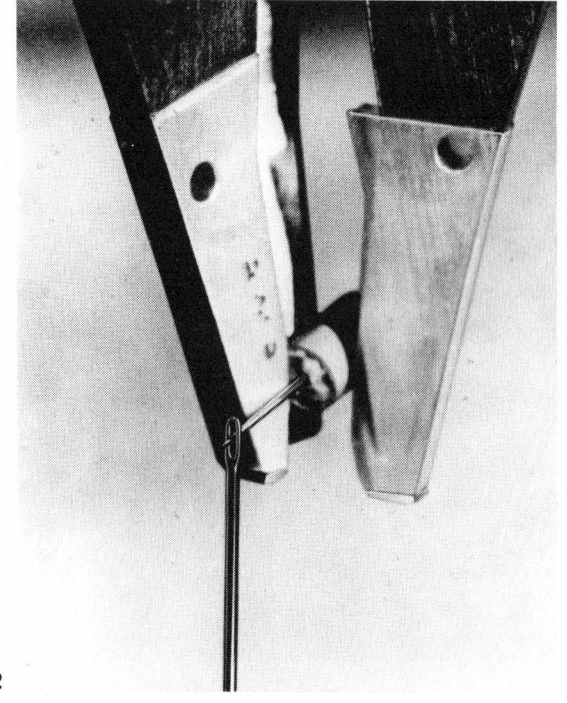

112

(113-127). Robots manufactured today can do herculean labor, while some give more modest performance. Yet even the budget efficient, less ambitious robot can spell the difference between profit and loss for a businessman. In ascending order of market value, here is a spectrum of what's available in robots these days: the $9,800 robot from International Robomation/Intelligence **(113);** Motion-Mate ($13,000) from Schrader Bellows **(114);** the Rhino Charger ($20,000) from Rhino Robots, Inc. **(115);** the Maker 100 ($40,000) from United States Robots **(116);** Unimate Puma Series 560 ($47,000) from Unimation **(117);** the DeVilbiss EPR 1000 ($55,000) from the DeVilbiss Company **(118);** the Limat 280 toggle arm arc welding robot ($59,000) from Comet Welding Systems **(119);** the ASEA IRB 6/2 ($60,200) from ASEA, Inc. **(120);** Unimate 2100 ($65,000) from Unimation **(121);** the DeVilbiss/Trallfa TR-3500 ($78,000) **(122);** T3-776 ($90,000) from Cincinnati Milacron **(123);** the DeVilbiss TR-4500 ($98,000 to $99,000) **(124);** the MTS 200A (under $100,000) from MTS Systems Corporation **(125);** the Nordson industrial coating robot (about $100,000 **(126);** the Limat 2000 welding robot ($131,000) from Comet Welding **(127).**

Photo 113 courtesy of International Robomation/Intelligence; photo 114 courtesy of Schrader Bellows; photo 115 courtesy of Rhino Robots, Inc.; photo 116 courtesy of United States Robots; photos 117 and 121 courtesy of Unimation, Inc., a Westinghouse Electric Company; photos 118, 122, and 124 courtesy of The DeVilbiss Company; photos 119 and 127 courtesy of Comet Welding Systems; photo 120 courtesy of ASEA, Inc. Industrial Robot Division; photo 123 courtesy of Cincinnati Milacron; photo 125 courtesy of MTS Systems Corporation; photo 126 courtesy of Nordson Corporation.

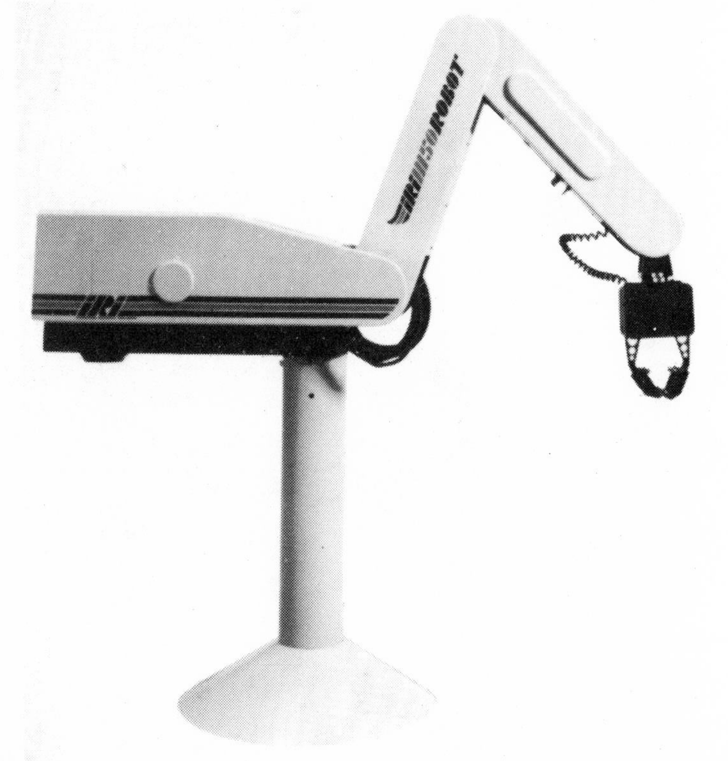

113

114

115

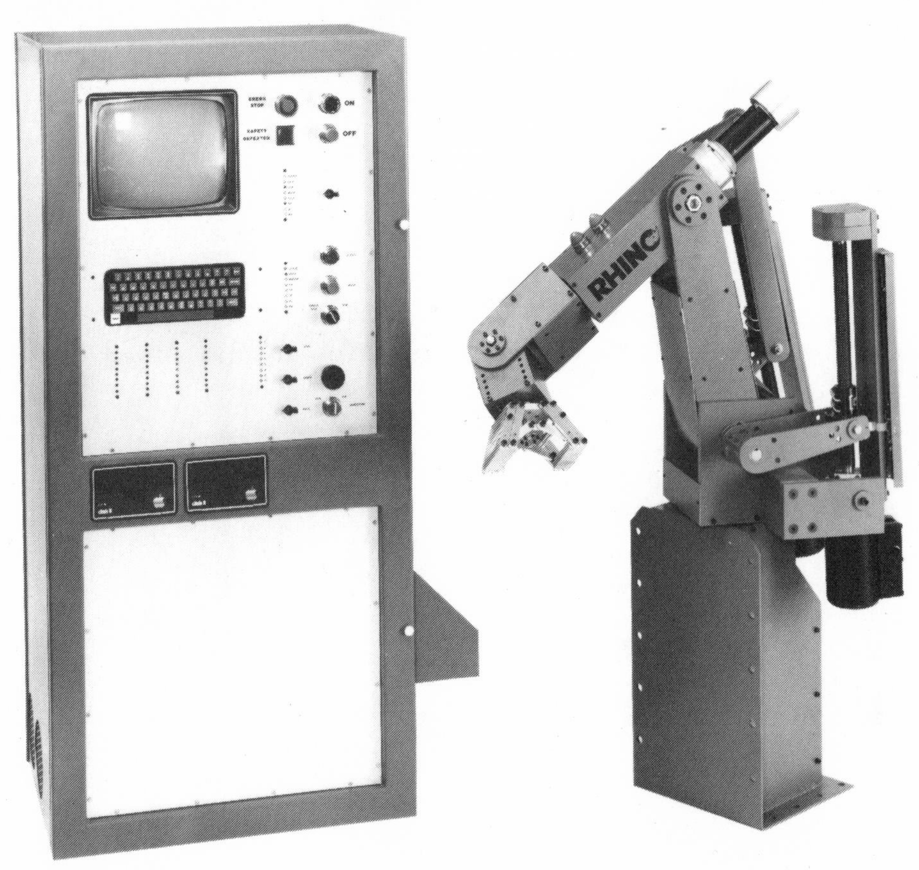

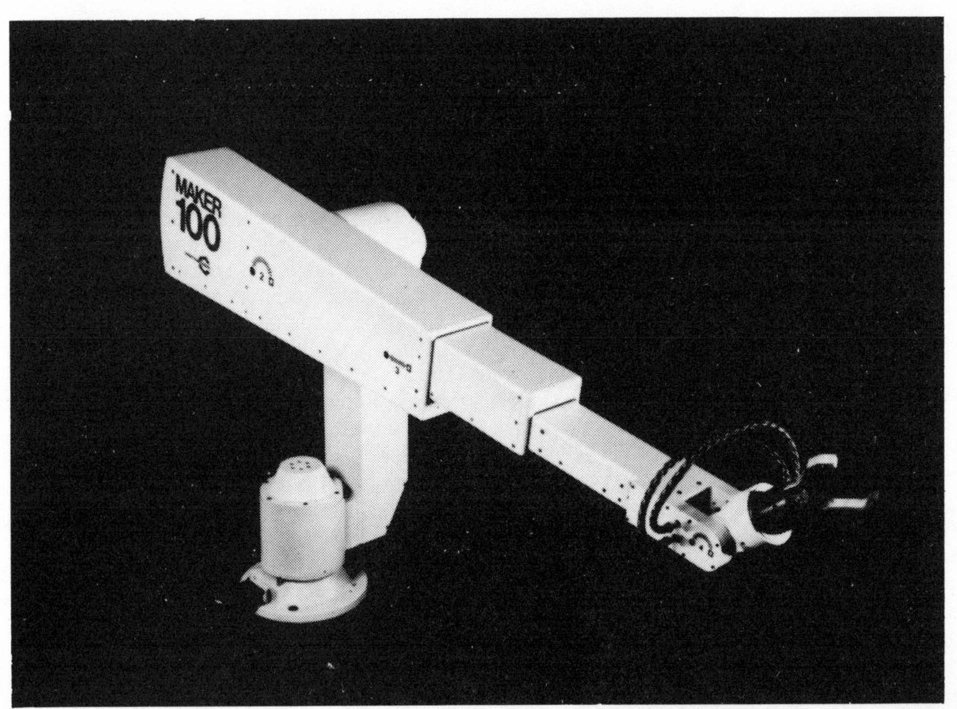

116

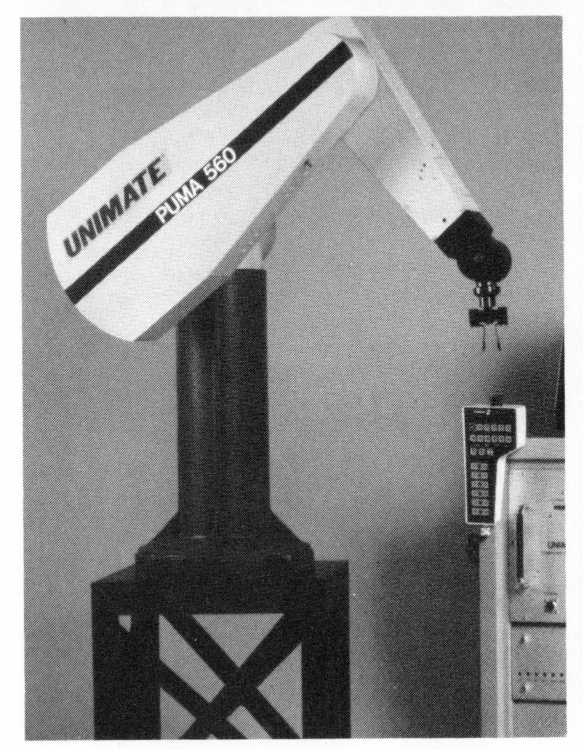

117

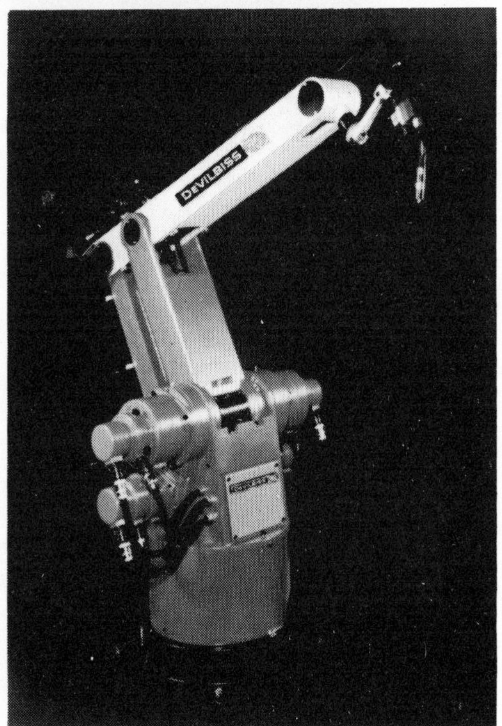

118

CONSUMER INFORMATION:

(113) International Robomation/ Intelligence announces: "YOUR $9,800.00 ROBOT IS READY TODAY." The IRI "affordable robot" is a material handling robot whose on-board printed circuit board contains eight computer systems. A robot control computer based on the 16-bit Motorola MC 68000 microprocesor is the central control computer in the hierarchy. Beneath the MC 68000 are six MC 6800 axis control computers which control and monitor the air motors that power the robot motion. A seventh MC 6800 computer serves as a safety computer which can shut the robot down automatically in the event of emergency. Several patents have been applied for in the design of the $9,800 robot. Also available to businessmen: a full-color 20″ × 30″ poster on IRI's $9,800 robot. To receive your free poster, or for complete ordering details, write on your company letterhead to Richard Carroll, Vice President of Sales, International Robomation/ Intelligence, 2281 Las Palmas Drive, Carlsbad, California 92008. Tel.: (619) 438-4424.

(114) Schrader Bellows Division of Scovill introduces MotionMate, a microprocessor-controlled programmable robot for under $13,000. MotionMate is used for light assembly applications, for loading and unloading machines, and for parts transfer from machine to machine. MotionMate is a five-axis, pneumatically powered robot that has a maximum payload of five pounds and can operate at speeds up to 24 inches per second. Programming is done on a hand-held teach module that uses simple graphic symbols for robot commands and requires no special skills of the operator. According to the manufacturer, such programming simplicity gives MotionMate the flexibility necessary for quick, easy changes to new production requirements. For more information, write for catalog MMI to Schrader Bellows Division, 200 West Exchange Street, Akron, Ohio 44309. Tel.: (216) 375-5202.

(115) The Rhino Charger is a six-axis robot that is geared for pick and place tasks and, according to Rhino Robots, Inc., is equally suited to industrial and educational locales: "The Charger, only 28 inches tall from floor to shoulder, is appropriate for a laboratory setting. With its twelve different types of hands, the Charger is capable of a wide variety of industrial tasks in addition to pick and place operations. These include feeding industrial presses, packaging bottles and cans, and feeding and loading conveyors." The Charger has a payload of up to fifty pounds. Price: $20,000. For additional information about the Charger contact Rhino Robots, Inc., 2505 South Neil Street, Champaign, Illinois 61820. Tel.: (217) 352-8485.

(116) At Robots 7 Exposition, three Maker 100 robots performed final assembly and testing of smoke detectors. A ceiling-mounted Maker 100 unloaded a smoke detector base from a box, placed it onto a fixture, and then rotated the base for proper orientation. The robot signaled the system controller to move the fixture to the second Maker 100 via a conveyor system. This robot installed a printed circuit board and a nine-volt battery, which is received from separate parts feeders. The battery was tested at the end of the battery feeding track, with all bad batteries rejected by the robot. After installing the board and a good battery, it signaled the system controller to move the fixture to the third Maker 100. This robot oriented, then picked up and placed the cap subassembly onto the base. It then actuated the test button on the top of the detector and, upon the system controller's verification of proper alarm intensity, sorted the detectors into pass or fail chutes. Price of an individual Maker 100: $40,000. For further information, contact United States Robots, 650 Park Avenue, King of Prussia, Pennsylvania 19406. Tel.: (215) 768-9210.

(117) The Unimate Puma Series 560 is a compact, high-speed electrically driven six-axis robot for assembly, finishing, and material handling. Features ±0.004 inch repeatability; five-pound payload. Price: $47,000. For more information, contact Unimation, Inc., Shelter Rock Lane, Danbury, Connecticut 06810. Tel.: (203) 744-1800.

(118) The DeVilbiss EPR-1000 is an electrically powered arc welding robot. It has been designed with special protective systems to accommodate load, overrun, and overcurrent protection. A double check system for overrun includes a software activated alarm stop and a limit switch stop. William D. Gauthier, director of DeVilbiss's Industrial Robot Operations, says: "In addition to this new electrically powered arc welding robot, DeVilbiss offers complete customer testing facilities." Price of the EPR-1000: $55,000. For more information, contact The DeVilbiss Company, Industrial Robot Operations, 837 Airport Boulevard, Ann Arbor, Michigan 48104. Tel.: (313) 668-6765.

(119) The Limat 280 toggle arm arc welding robot can be mounted for use either standing in an upright position or, if required, a suspended position. The welding torch is attached so that there is no restriction or interference with incoming cables. The manufacturer, Comet Welding Systems, claims: "The exclusively designed main axis and the speed of movement create a robot working area much larger than competitive systems." Price: $59,000. For more information, contact Comet Welding Systems, 800 Nicholas Blvd, Elk Grove Village, Illinois 60007. Tel.: (312) 956-8717.

(120) The ASEA IRB 6/2 is shown applying structural adhesive to an automotive panel. The adhesive bonding system, based on the ASEA robot and its SII controller, uses Motorola 6800 16-bit microprocessor. According to ASEA, Inc., the manufacturer of the robot, the IRB 6/2 can "effectively double or triple the productivity of manually operated adhesive bonding station." The robot can be used with urethane sealants, two-component epoxy resins, and the newer, high-shear strength structural hot-melt adhesives. Price: $60,200. For further information, contact ASEA, Inc., 4 New King St., White Plains, N.Y. 10604. Tel.: (914) 428-6000.

119

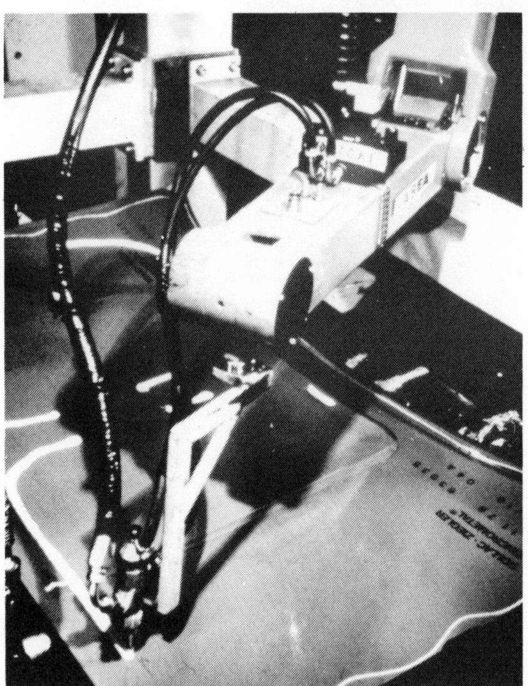

120

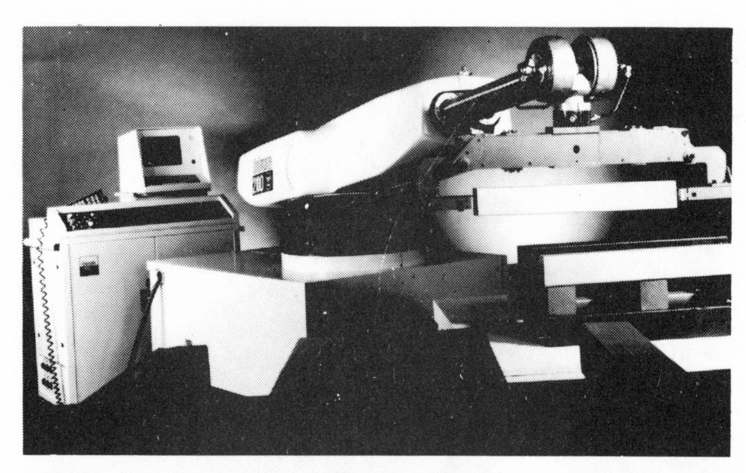

121

122

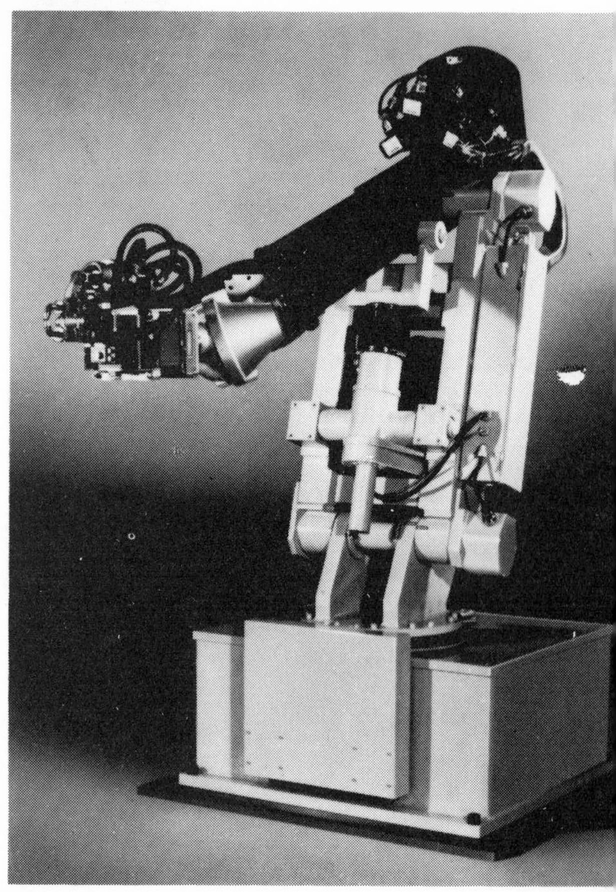

123

(121) Unimation Inc. calls its series Unimate 2100 "the most widely used industrial robot." Applications for the VAL computer controlled robot include material handling operations, arc and resistance welding, palletizing, investment casting, press loading, and machine loading. The 2100 has five degrees of motion, a payload capacity of 175 pounds depending upon operating speed, and repeatability of 0.05 inches. Price: $65,000, with control system. For more information, contact Unimation, Inc., Shelter Rock Lane, Danbury, Connecticut 06810. Tel.: (203) 744-1800.

(122) The DeVilbiss/Trallfa TR-3500 finishing robot is shown spray painting metal shutters. The manufacturer states: "Right now, more than one thousand of our systems are at work around the world helping small and large companies alike achieve maximum productivity in their finishing operations." Schools are available to DeVilbiss/Trallfa users to train maintenance personnel in the practical techniques required to maintain the units. Service contracts can be arranged. Component repair and service parts are available from the company's Ann Arbor, Michigan, manufacturing facility. Emergency service personnel and parts are normally available within twenty-four hours or less. Price of robot: $78,000. For more information, write to The DeVilbiss Company, PO Box 913, Toledo, Ohio 43692. Tel.: (419) 470-2169.

(123) Cincinnati Milacron's T^3-776 robot can be used in a number of operations, though the manufacturer says spot welding is its primary application. The robot has six axes, with three of the axes contained in Milacron's patented three-roll wrist. According to Cincinnati Milacron: "The flexibility of the three-roll wrist enables the robot to move the weld gun in and around tight, difficult to reach areas quickly and accurately for fast, precise placement of the spot welds." Price: $90,000, fully equipped. For further information, contact Cincinnati Milacron, Industrial Robot Division, Lebanon, Ohio 45036. Tel.: (513) 932-4400.

(124) DeVilbiss calls its TR-4500 its most sophisticated spray-finishing robot. The manufacturer claims: ". . . DeVilbiss/ Trallfa robots are the number one selling finishing robot in the U.S., with over 1,200 units in operation worldwide." Price: $98,000 to $99,000. For further information, contact The DeVilbiss Company, 300 Phillips Avenue, Toledo, Ohio 43692. Tel.: (419) 470-2169.

(125) The MTS 200A has been applied for parts handling, tool manipulation, palletizing, testing and inspection, machine loading and unloading. It is capable of rapidly moving 220-pound loads at speeds up to 150 inches per second. Price: under $100,000. For further information, contact MTS Systems Corporation, Industrial Systems Division, Box 24012, Minneapolis, Minnesota 55424. Tel.: (612) 937-4000.

(126) The Nordson industrial coating robot is routinely used to spray paint. It can also be used, however, to apply waxes, sealants, sound-deadening compounds, and various other coating materials. The Robotics Division of Nordson Corporation offers a new field service to potential buyers: a robot application evaluator for simulating and assessing the capabilities of the Nordson robot on the actual production line. This means that an industrial finisher interested in coating robot applications can test the feasibility on location without installing an actual robot system. The user can test spray his products to see if a robot will satisfy his particular application needs. Because the part is actually coated during the test, it can be evaluated for coverage and overall quality as well. The simulator arm is self-contained for rapid assembly and can be used either with the customer's existing paint system or a portable Nordson electrostatic system. Price of the robot: about $100,000. For further information, contact Nordson Corporation, Robotics Division, 555 Jackson Street, PO Box 141, Amherst, Ohio 44001. Tel.: (216) 988-9411.

(127) The Limat 2000 welding robot

124

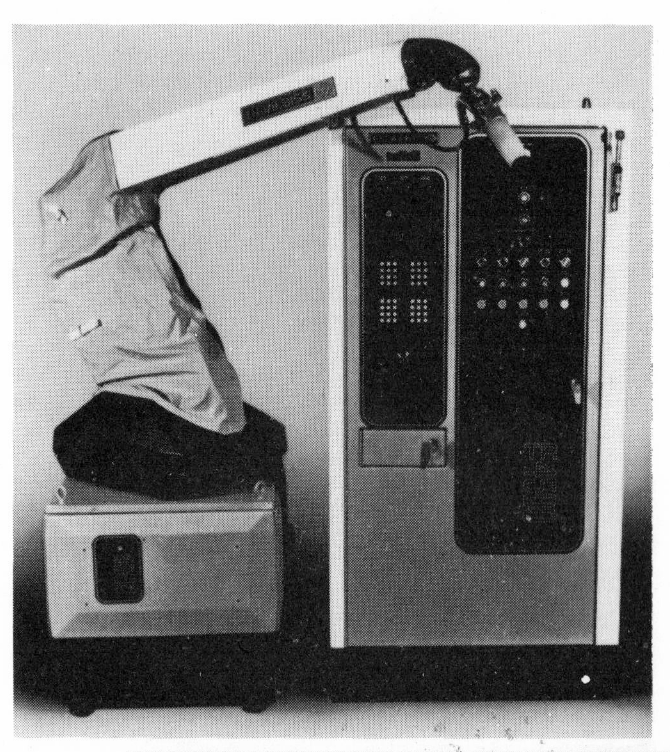

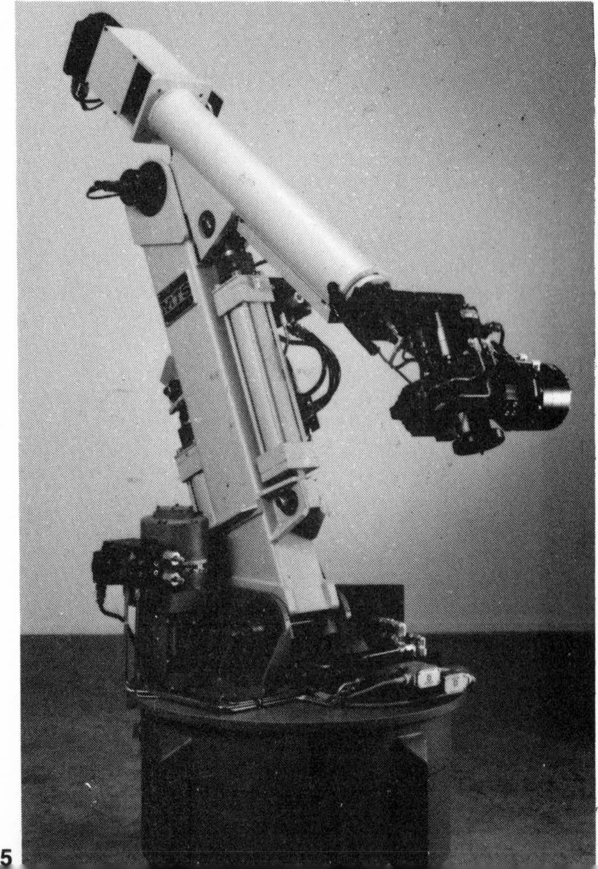

125

features two independent welding heads that can be programmed to operate separately or in concert. A portable teach-in programing board controls operation through a Digital PDP 11/03 microcomputer with 32K memory. The precision of the computer and the construction of the robot combine to offer a weld accuracy of $\pm$ 0.1 mm (0.004 inches). Price: $131,000. For further information, contact Comet Welding Systems, 800 Nicholas Blvd, Elk Grove Village, Illinois 60007. Tel.: (312) 956-8717.

126

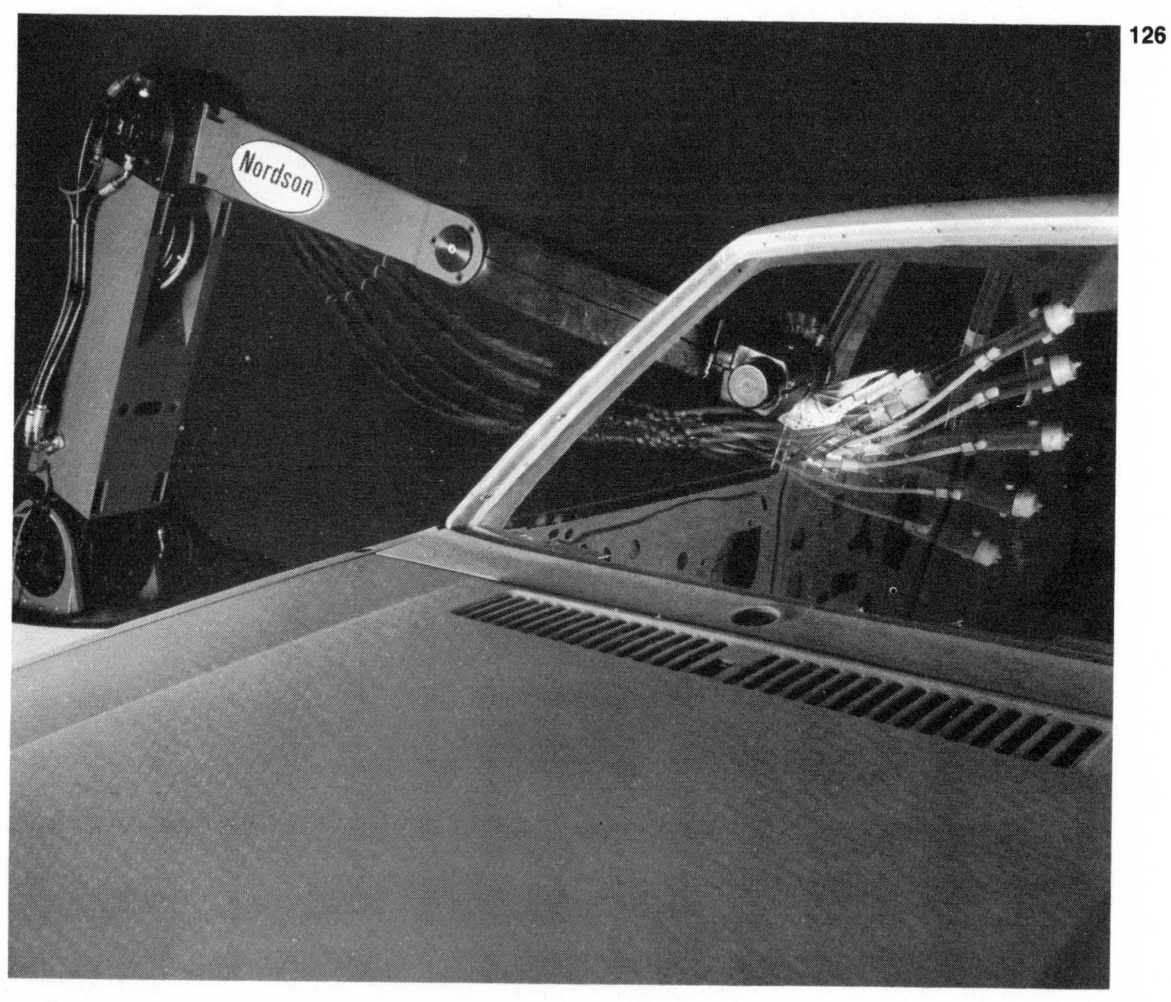

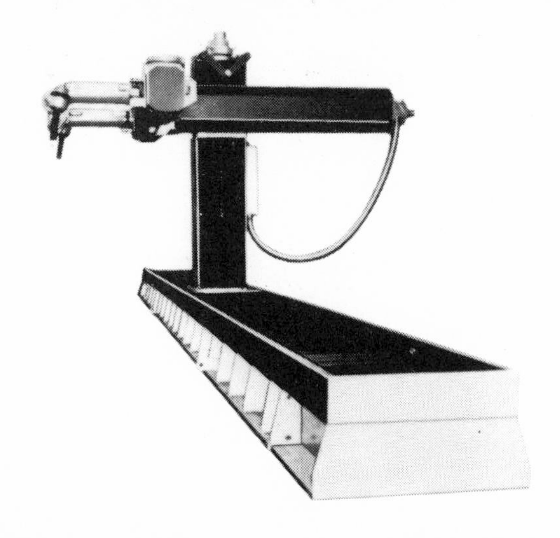

127

(128-129)
Control Automation's MiniSembler **(128)** and Intelledex's 605 T **(129)** are light assembly robots designed for use in high technology manufacturing. *Photo of MiniSembler, courtesy of Control Automation Inc.; photo of 605 T, courtesy of Intelledex Incorporated.*

CONSUMER INFORMATION

According to its manufacturer, "The MiniSembler is intended primarily for high precision assembly of printed circuits, calculators, small motors, disk drives, and keyboards." The MiniSembler is a benchtop sized system with .0001-inch precision. Designed as a computer peripheral, the robot can be used with any computer programmed in any standard computer language and can be networked with other robots, sensors, or factory management systems (either as a single work station or as part of a complete assembly work cell). Price: $37,000.

Of its 605 T, Intelledex states that it features ".002 inch accuracy per foot from the calibrated work surface." Price: $48,000.

For further information on the MiniSembler contact Control Automation, Inc., Princeton-Windsor Industrial Park, PO Box 2304, Princeton, N.J. 08540. Tel.: (609) 799-6026. For further information on Intelledex's 605 T contact Intelledex Incorporated, 33840 Eastgate Circle, Corvallis, Oregon 97333. Tel.: (503) 758-4700.

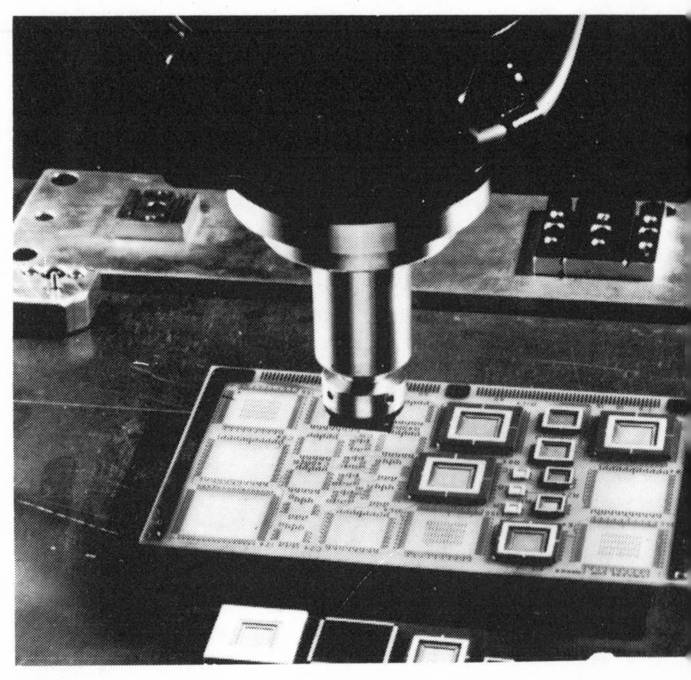

128

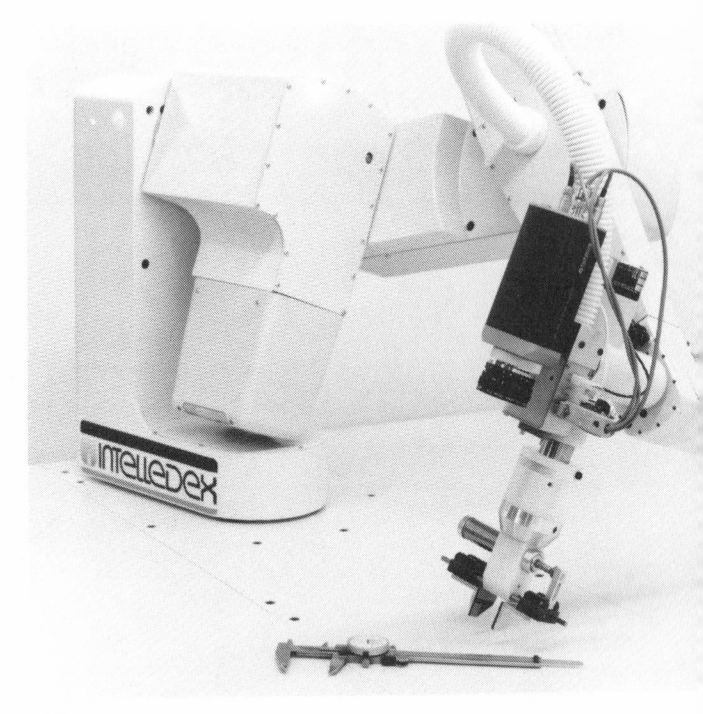

129

A sighted and touch-sensitive parallel jaws gripper from Object Recognition Systems **(130)** and a vision system called Autovision from Automatix, Inc. **(131)** represent refinements in sensors that increase the capacities of industrial robots. Vision systems, in particular—such as these shown at General Motors facilities **(132-133)**—are capable of sophisticated operations. *Photo 130 courtesy of Object Recognition Systems, Inc.; photo 131 courtesy of Automatix; photos 132 and 133 courtesy of General Motors Corporation, Detroit, Mich.*

CONSUMER INFORMATION:
The parallel jaws gripper from Object Recognition Systems, Inc., serves as the hand for an industrial robot arm. The gripper is equipped with pressure and optical sensors so that it can retrieve delicate parts. For further information, contact Object Recognition Systems, Inc., 1101-B State Road, Princeton, N.J. 08540. Tel.: (609) 924-1667.

Automatix, Inc. has several vision systems specially adapted for different needs. Autovision is a programmable image sensing and processing system that inspects, identifies, counts, sorts, positions, and orients parts data. Prices for the programmable vision system average $35,000. Delivery is from 30 to 90 days ARO. For further information on Automatix products, contact Automatix Inc., 1000 Tech Park Drive, Billerica, Massachusetts 01821. Tel.: (617) 667-7900.

131

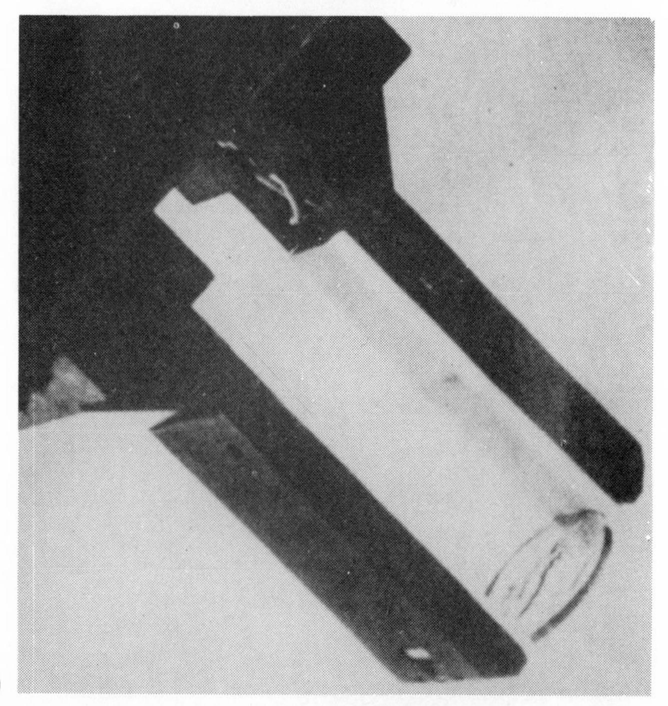

130

A sense of how these vision systems operate is gained through photos of operations at General Motors facilities. In the laboratory demonstration shown here **(132)**, an employee of the Computer Science Department is seen placing a part on a conveyor belt. A television camera mounted in the apparatus at the rear views the moving conveyor belt below it and the computer forms a digitized picture of the part (a representative picture is shown in the inset). The computer goes through a series of recognition techniques and, when it makes a match with a similar part already in its program, advises the robot arm where the part will be on the belt so that it may pick it up. In testing at GM Manufacturing Development, GM Technical Center in Warren, Michigan **(133)**, four major elements of the system are visible: the robot unit and its control system, including specially modified software to permit acceptance of visual commands; the vision element consisting of the light system ("line of light"—inset at top left), camera, and computer, where the image is stored until used; a motion sensor or encoder on the conveyor belt to keep track of the parts after they've been seen by the camera; and the supervisory computer which takes all of the information provided by the camera (shape, size, location, and orientation of parts travelling on the belt), the belt encoder, and the robot and then directs the robot element to pick up the parts and properly dispose of them, according to a preprogrammed scheme. A system such as this would lend itself to material handling, transferring and packaging or palletizing of larger, heavier parts.

132

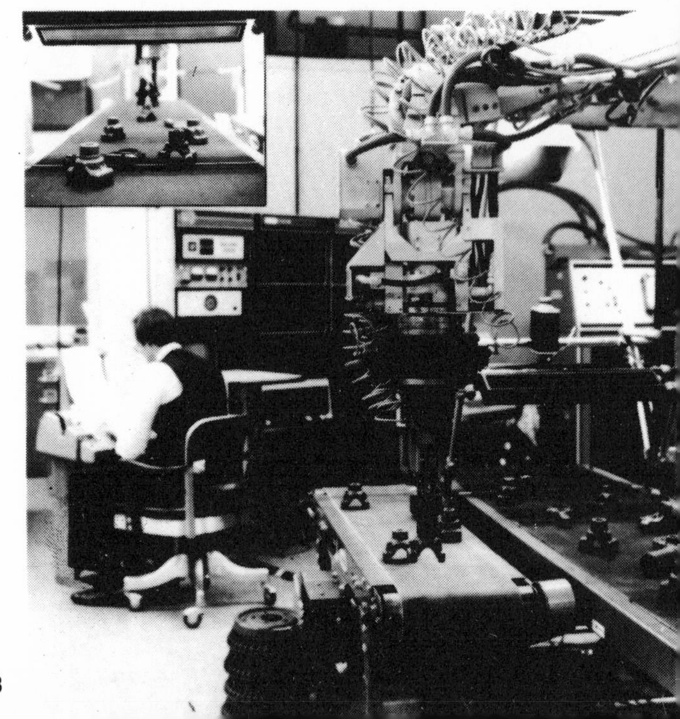

133

(134) A GE P5 process robot, coupled to GE's advanced vision system, welds an irregularly shaped joint. *Photo courtesy of General Electric Company.*

CONSUMER INFORMATION:
In GE's advanced vision system, the TV set in the foreground shows the scene observed by the system's "eyes" (which are located in the welding torch assembly). The two vertical lines are laser stripes that are beamed across the weld joint (the dark horizontal line) as a navigational aid. The bright area shaped like a reversed letter C is a part of the welding arc, and the weld puddle trails to the right. (The reversed C image is created by a shielding device that blocks out some of the intensely bright arc to keep it from "blinding" the vision sensor.) GE calls its intelligent vision welding system WeldVision. The WeldVision system includes: WeldVision Intelligent Vision Processor, including a GE 2150 Solid State Buffer

134

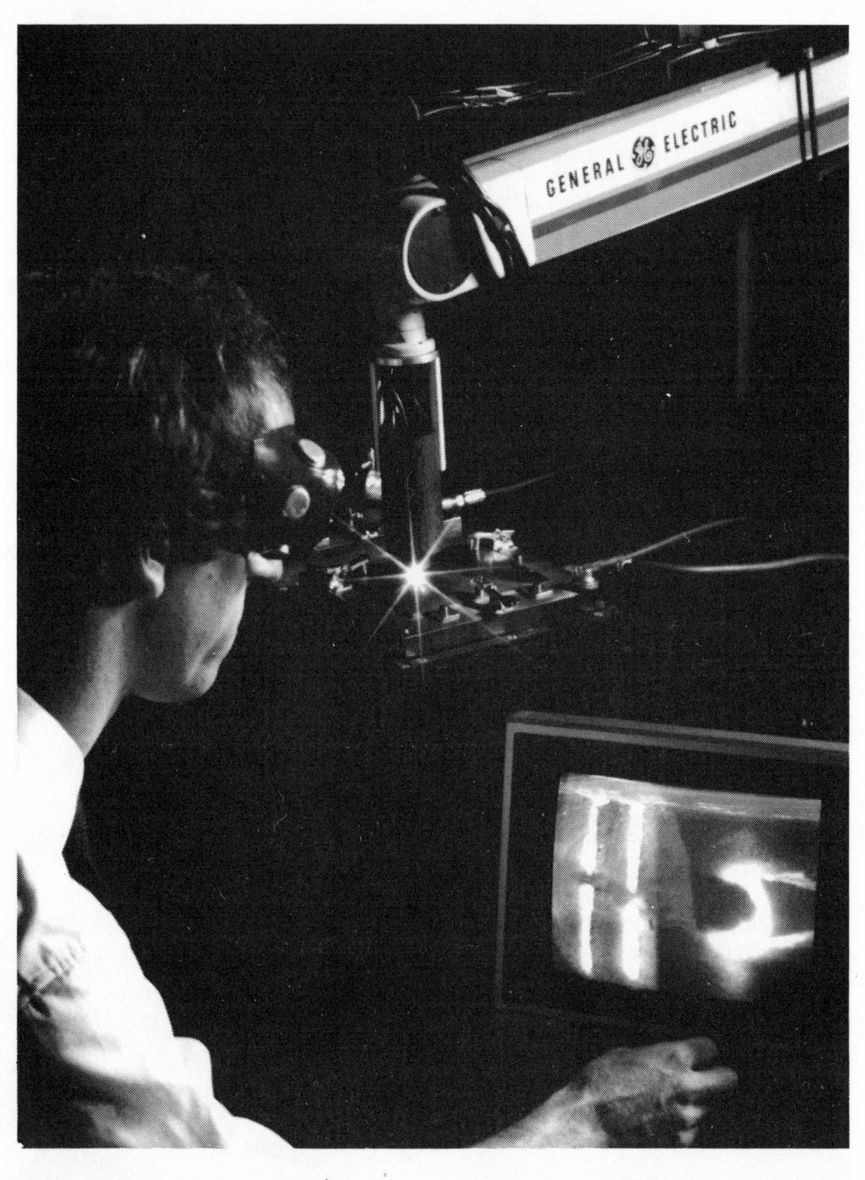

135

Memory and a state-of-the-art multiprocessor controller; GE Solid State CID Camera; Coherent fiber optic bundles for through-the-torch imaging; optical and electrical filtering. List price for the WeldVision System: $24,500. The GE Robotic TIG (tungsten inert gas) Welding System with WeldVision: $112,000. For more information about robot systems, intelligent vision systems, robot application consulting and robot training, GE has a toll-free number: (800) 626-2001, ext. 333.

For background on vision systems, also available: A three-volume report (350 pages) from Technical Insights, Inc.—*Machine Vision for Robotics and Automated Inspection.* Price: $185. Overseas buyers, add $25. For order form, write Technical Insights, Inc, PO Box 1304, Ft. Lee, N.J. 07024.

(135) A flexible assembly (incorporating robots) from Design Technology Corporation. *Photo courtesy of Design Technology Corporation.*

CONSUMER INFORMATION:
Robotic assembly lines have been used overseas for some time. In this country, Design Technology Corporation has announced that it will design and build special automated lines which integrate robots with conventional parts handling devices and conveyors. Design Technology will use robots from leading robot companies all over the world to provide custom-built lines tailored to the exact needs of the user. The system depicted above assembles a family of seven different automotive products at 2,400 assemblies per hour. According to Design Technology, the new system cuts assembly labor by more than 80 percent while productivity is increased. For further information, contact Design Technology Corporation, Second Avenue, Burlington, Massachusetts 01803. Tel.: (617) 272-8890.

(136-137) Unimation robots now bear the Westinghouse insignia—see the circled "W" on the Unimate 900 (price: $25,000)—since being bought by Westinghouse Electric Corporation **(136)**. Westinghouse is not alone among major corporations moving in on the robot market. For example, IBM—whose robot is seen working an EWAB conveyor belt **(137)**—is one of several heavyweight newcomers. *Photo 136 courtesy of Unimation, Inc., a Westinghouse Electric Company; and 137 courtesy of EWAB Engineering, Inc.*

136

137

138

(138-139)
High technology throughout Chrysler Corporation's Jefferson Assembly Plant in Detroit is typified by this robot welding line, which is computer controlled and programmed to apply nearly 3,000 welds to each body cycled through the system **(138).** In the other photo, GCA's XR6050 robot is shown. The XR6050, among other applications, "listens" for leaks in an automobile door frame with an ultrasonic probe **(139).** *Photo 138 courtesy of Chrysler Corporation; and 139 courtesy of GCA/Industrial Systems Group.*

CONSUMER INFORMATION:
Price for GCA's XR6050: about $100,000. For information, contact GCA Corporation, 209 Burlington Road, Bedford, Mass. 01730. Tel.: (617) 275-9000.

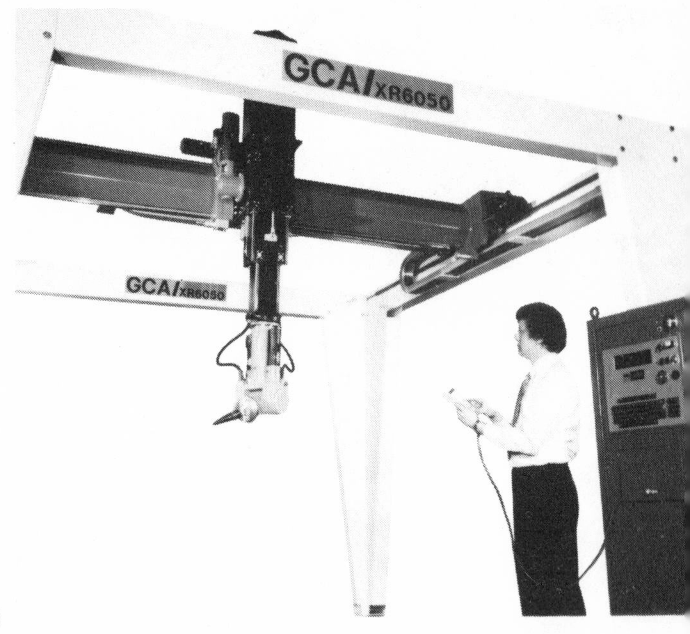

139

(140) Bomb carrier robot. *Photo by UPI.*

CONSUMER INFORMATION:
For further information, contact Pedsco Canada, Ltd., 180 Finchden Square, Unit 3, Scarborough, Ontario, Canada M1X 1A8. Tel.: (416) 298-9989.

(141) Herman: a robot used at toxic or radioactive sites. *Photo courtesy of Nuclear Division, Union Carbide Corporation.*

CONSUMER INFORMATION:
The protocol for requesting Herman's assistance involves contacting Hershel Hickman of the Department of Energy, PO Box E, Oak Ridge, Tennessee 37830. Tel.: (615) 576-0753. With Hickman's okay, the robot and operators are dispatched to the troubled site. Union Carbide receives what it terms a "full recovery" fee—money that covers the salaries of the robot's personnel, transportation, lodgings, and meals. Union Carbide does not sell the Y-12 plant mobile manipulator, as Herman is known. It paid $63,010 when it bought the robot several years ago. For details on purchasing Herman, contact Programs and Remote Systems Corporation, 3460 Lexington Avenue North, St. Paul, Minnesota 55112. Tel.: (612) 484-7261.

140

141

(142) A life-scale model of the Viking Lander used to explore the surface of Mars. *Photo by NASA.*

CONSUMER INFORMATION:
The model of Viking Lander is kept at the NASA Jet Propulsion Laboratory. The Lander's dimensions: about 5 feet across and $1\frac{1}{2}$ feet high; it weighs 1 ton. Its soil sample collecting arm extends about 10 feet to the lower right. According to NASA: "The cameras are the vertical cylinders, each with a vertical black slit. The disk, top rear, is the S-band high-gain radio antenna that transmits to Earth the camera pictures and scientific data."

(143) A view from the Viking Lander, Mars. *Photo by NASA.*

CONSUMER INFORMATION:
Toward the left in this panorama of the Martian surface is a small dune of fine-grained material scarred by trenches dug by Viking I's surface sampler in 1976. The sampler scoop is seen in its temporary parked position. When in motion, soil was collected for the gas chromatograph mass spectrometer, the instrument that analyzed the surface material for the presence of organic molecules. NASA photos like these are sold to the general public. For a list of available color and black and white prints, write to Les Gaver, Chief, Audiovisual Section, National Aeronautics and Space Administration, Washington, D.C. 20546. Photos come in different sizes and are priced accordingly. Example: a black and white 8 × 10 print costs $2.50. Gaver says: "Because of budget limitations, we make no free distribution of our photos except to information media representatives. However, prints and/or transparencies are available at your expense." The catalog of official NASA photographs contains an order blank.

143

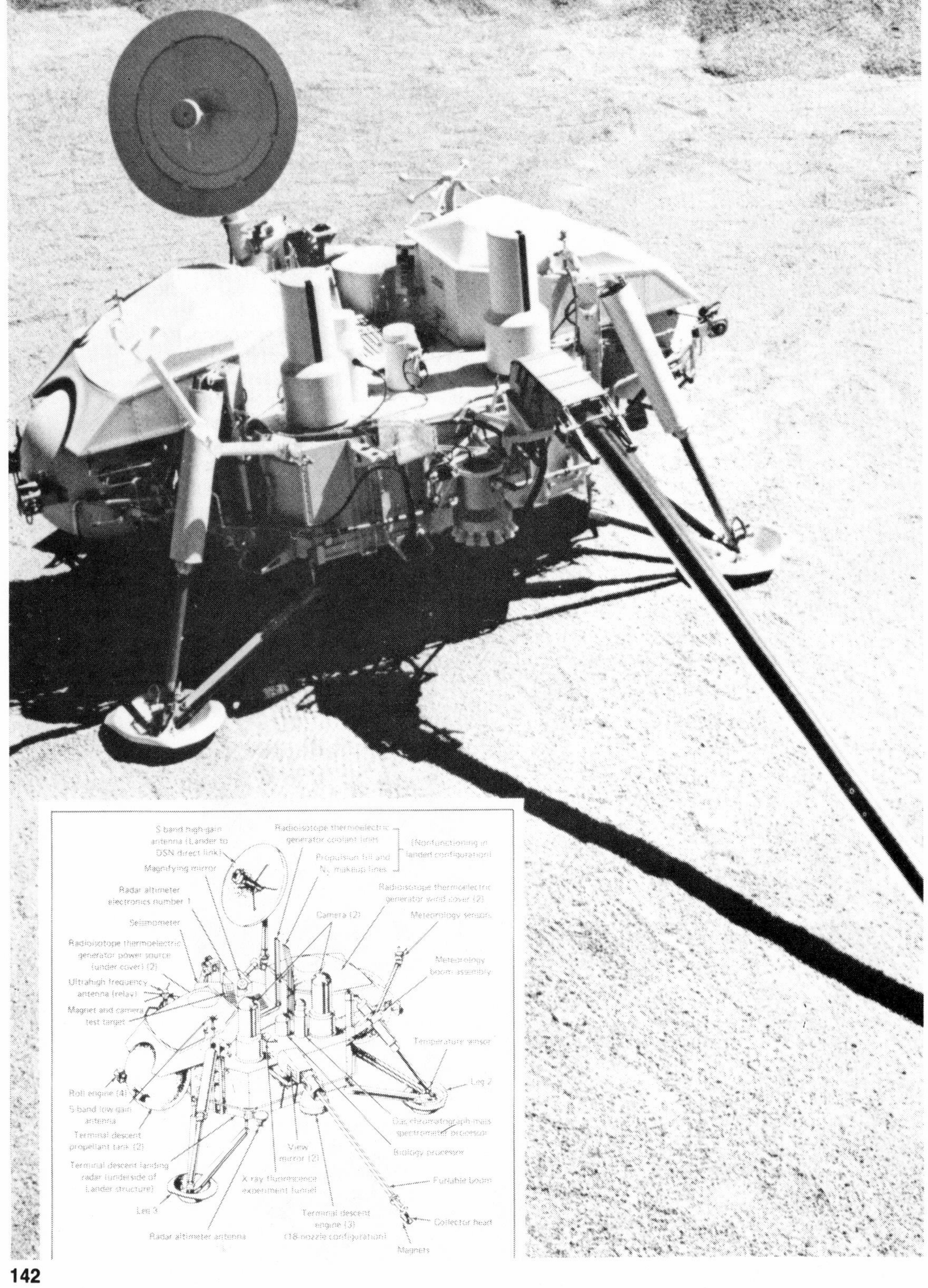

S band high-gain antenna (Lander to DSN direct link)
Radioisotope thermoelectric generator coolant lines
Propulsion fill and N_2 makeup lines
(Nonfunctioning in landed configuration)
Magnifying mirror
Radar altimeter electronics number 1
Radioisotope thermoelectric generator wind cover (2)
Camera (2)
Meteorology sensors
Seismometer
Radioisotope thermoelectric generator power source (under cover) (2)
Meteorology boom assembly
Ultrahigh frequency antenna (relay)
Magnet and camera test target
Temperature sensor
Leg 2
Roll engine (4)
S band low-gain antenna
Terminal descent propellant tank (2)
Gas chromatograph mass spectrometer processor
View mirror (2)
Biology processor
Terminal descent landing radar (underside of Lander structure)
X ray fluorescence experiment funnel
Furlable boom
Leg 3
Terminal descent engine (3) (18-nozzle configuration)
Collector head
Radar altimeter antenna
Magnets

(144) The Proximity Operations Vehicle: on the drawing board. *Photo courtesy of Grumman Aerospace Corporation.*

(145) The Mailmobile delivery vehicle, from Bell & Howell. *Photo provided by Bell & Howell, Automated Systems Division, Zeeland, Michigan.*

CONSUMER INFORMATION:
The Mailmobile is a self-propelled battery-driven delivery vehicle for modern offices. It is guided by an invisible chemical line that only it can see. The robot vehicle automatically stops at designated points along its path to pick up and deliver mail, word and data processing materials, photocopies, office supplies, and more. According to Bell & Howell: "The guidepath is applied over existing floor covering and does not require alteration of building content or structure. Length of guidepath is unlimited. . . . The Mailmobile detects two types of stops—temporary (enroute distribution stops) and permanent (end of route). The stop indicators are also essentially invisible. When a temporary stop is detected, the Mailmobile pauses for a preset time (4, 12, 20, or 28 seconds) and then resumes normal operation. Number of temporary stops is unlimited. Detection of a permanent stop halts the Mailmobile until manually restarted." Mailmobile is 24 inches wide, 58 inches long (69 inches with rear container attached), 51 inches tall; it weighs 650 pounds and is made of aluminum with baked vinyl epoxy finish. Its capacity is 800 pounds on a level surface. Automatic operation occurs only on guidepath. Loss of guidepath causes immediate application of full dynamic braking. A beep tone is emitted during automatic travel, beginning two seconds prior to movement. Volume control is provided. An internal chime (optional) sounds a distinctive tone at each temporary stop. Blue running lights flash alternately during automatic operation. Lights flash faster when stopped for obstacles, or loss of guidepath. Price: $20,000. For more information contact Bell & Howell, Automated Systems Division, 280 E. Riley Street, Zeeland, Michigan 49464. Tel.: (617) 772-1000.

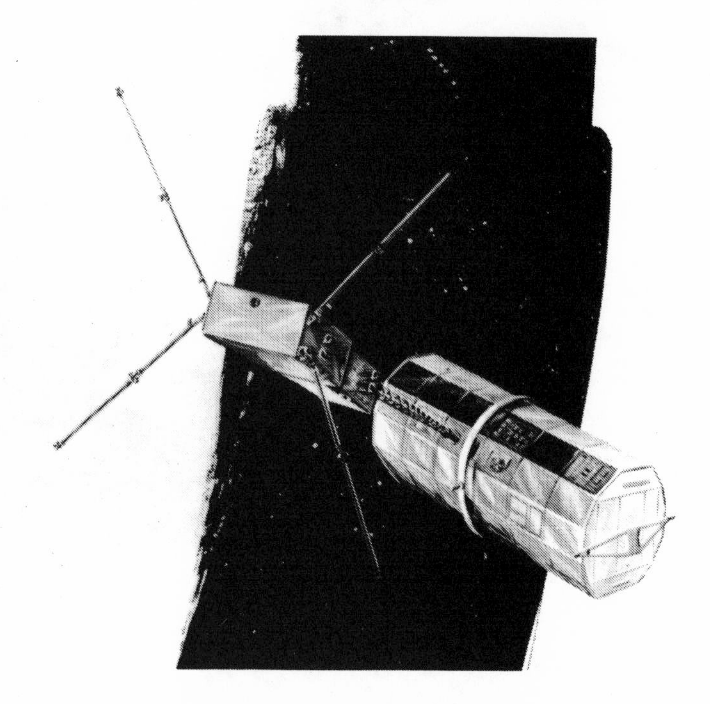

144

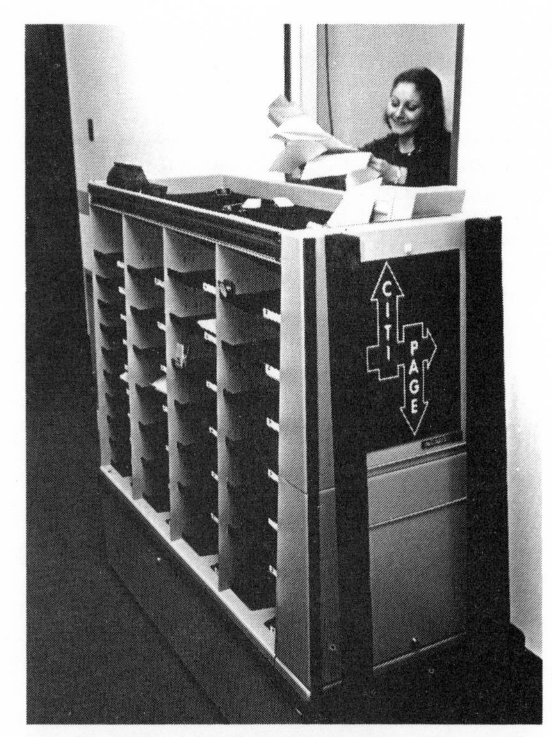

145

(146) The Zymate. *Photo courtesy of Zymark.*

CONSUMER INFORMATION:
The Zymate Laboratory Automation System (from Zymark) combines robotics and state-of-the-art microprocessor technology to perform common laboratory operations used in sample preparation. According to its manufacturer: "A Zymate System will weigh, dilute, mix, and transfer samples to test tubes or vials for any chemical sample preparation procedure; homogenize, centrigue, and extract samples for biological testing; pipet, filter, coentrate, and derivatize samples for chromatography or spectroscopy." The Zymate System consists of Zymate controller with user memory, laboratory robot, and general purpose hand with holder for automatic hand changing. Includes EasyLab Application Software and floppy disk (5¼-inch) drive for storing and retrieving sample preparation methods. The Zymate Controller can operate up to twenty-five laboratory stations. Price: $18,500. For more information, contact Zymark Corporation, Zymark Center, Hopkinton, Massachusetts 01748. Tel.: (617) 435-9041.

146

(147) Sheep-shearing robot from Ex-Cell-O Corporation. *Photo by Phil Berger, courtesy of Ex-Cell-O Corporation.*

CONSUMER INFORMATION:
At Robots 7 Exposition in Chicago (1983), there stood a blinking toy robot at the Ex-Cell-O booth next to a sign that said: "Hello, I'm Robbie. I'm celebrating my 20th anniversary. If you look over my shoulder, you'll see Ex-Cell-O's newest robotic arm. This unique arm is designed to painlessly shear sheep." According to the Ex-Cell-O spokesman there, the robot on display—replete with electric shear at the end of the robotic arm—was a prototype sold to a company in Australia, where there are millions of sheep and a need for such a robot. That need, the spokesman said, is not significant enough elsewhere for marketing a sheep-shearing robot. At the time, Ex-Cell-O was not quoting a price for its robot prototype. For further information, contact Ex-Cell-O Corporation, 525 Berne Street, Berne, Indiana 46711. Tel.: (219) 589-2136.

(148) This maintenance manipulator is a remote controlled device to repair deep sea oil-producing equipment used by Exxon. For further information, contact Exxon Company USA, PO Box 60626, New Orleans, Louisiana 70160. Tel.: (504) 561-3636. *Photo courtesy of Exxon Company, U.S.A.*

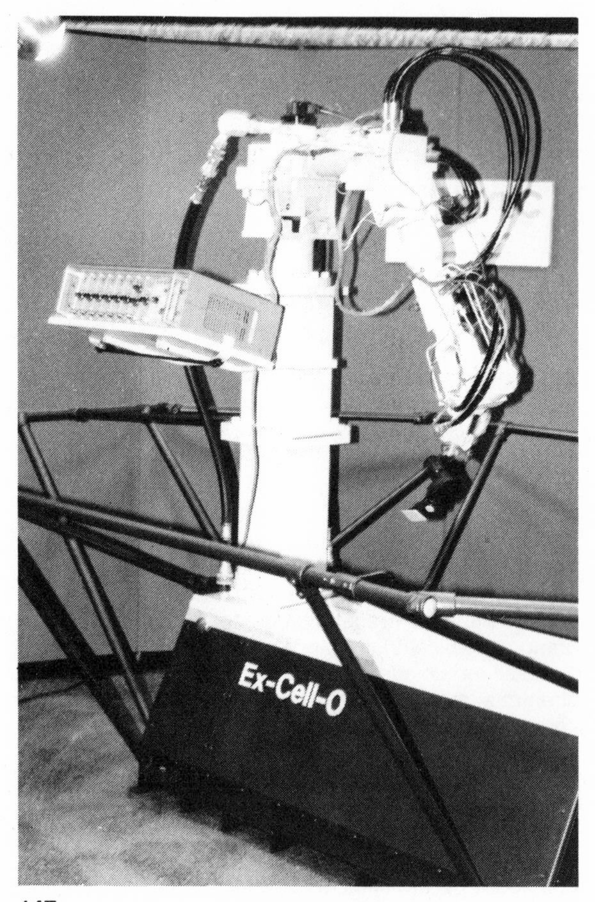

147

148

(149-150) Two uses for the medical manipulator—a robot-like arm that expands the functions of a quadriplegic. Medical manipulator is a prototype being developed by Carl P. Mason for the Veterans Administration Prosthetic Center. For further information, contact Director, Veterans Administration Prosthetic Center, 252 Seventh Avenue, New York, New York 10001, Tel.: (212) 620-6702. *Official Veterans Administration Photographs.*

(151) The AbilityPhone terminal not only serves as an accessible telephone for the handicapped, but it can also perform essential household tasks. Suggested retail price: $2,335. For further information, contact Basic Telecommunications Corporation, 4414 East Harmony Road, Fort Collins, Colorado 80525. Tel.: (303) 226-4688. *Photo courtesy of Basic Telecommunications.*

149

150

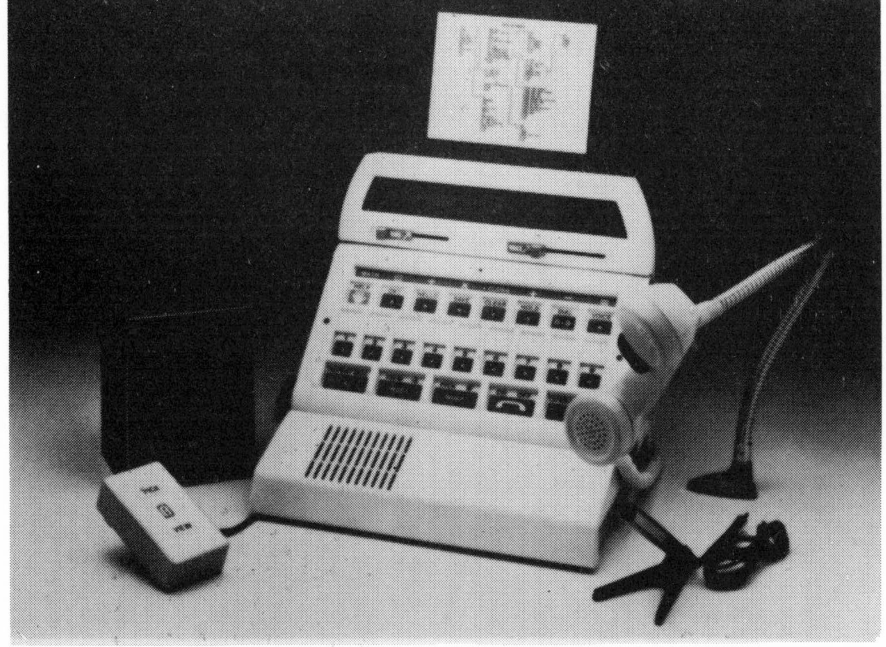

151

THE FUTURE ROBOT

What's ahead in robotics

THE FUTURE ROBOT

If robots were a whimsical notion not so long ago, more properly the province of speculative writers, that day clearly is behind us.

Robots are now part of our lives, their usages increasing all the time. For futurists, the role of robots has barely been tapped.

According to a report of the Japan Industrial Robot Association: "From 1985 to 1990, robots will come into use on a massive scale in spraying insecticide on farms, spreading fertilizer, inspecting eggs and packing them, milking cows, cutting lumber, in planting and in ocean research."

The Trade and Industry Ministry in Japan has researchers working on developing so-called super robots to assume dangerous jobs that humans now perform. The objective is to design and build robots that will fight fires, maintain a nuclear power plant, or work in the depths of the ocean. Projected cost for the eight-year program? More than $85 million.

Through robotic ingenuity, nations with limited resources will compete with the industrial powers. Take the case of Israel, as reported by Sheila Evan-Tov: "The Robotics Laboratory at the Technion in Haifa is *the* source of Israel's robotic industry. . . . All the industrial robots on the drawing boards for Israeli industry have their genesis at the Technion. These include a device that assembles twenty components into a stove-top range, a 'robot doctor' that diagnoses the robots' ills and instructs a technician how to cure them. . . . Eventually, says professor [Yoram] Koren [Senior Lecturer at Technion's Faculty of Mechanical Engineering and head of the Robotic Laboratory], oranges and avocadoes grown in Israel will be picked and packaged by mechanical hands. Completely automatic factories will design and manufacture products by computer. 'All phases of production, assembly, and testing will be robot-operated,' predicts Koren, 'with the minimum of human intervention.' "

From *The New York Times*, July 3, 1983: "Grocery store robots? Why not, figures the Seiyu [Japan] supermarket chain. . . . The company plans to install a computer-controlled system using carrier robots to stock a full store by moving the goods out of the storeroom to the appropriate aisle."

The technical precision of today's robots—like Merlin's needle-threading ability or the tour de force performed by IBM's 7565 of gently picking up an egg and, on cue, breaking it—is already impressive. But roboticists say that the robot of the future may not need to be so accurate but rather have the capacity to adapt, as humans do, to the imprecise factory milieu. And who among us—given the rapid progress these roboticists already have made—doubt that tomorrow's robots won't be up to the job?

INDEX OF ROBOTS

Page numbers in *italics* indicate illustrations or captions.

INDEX OF MANUFACTURERS

Page numbers in *italics* indicate illustrations or captions.

SUBJECT INDEX

Page numbers in *italics* indicate illustrations or captions.